Android 移动应用开发项目实战教程

主　编　路　景　赵志成
副主编　翟政凯
参　编　张诚洁　高　伟

北京理工大学出版社
BEIJING INSTITUTE OF TECHNOLOGY PRESS

内容简介

本教材以实战项目——"学习通关"App 设计与开发为载体，在项目开发的推进中贯穿 Android 移动应用开发的相关知识和技术。全书分为 6 个部分：项目导读、搭建 Android 开发环境、用户管理功能的实现、限时答题功能的实现、学习模块的实现、电台模块的实现。

本书通过真实的项目开发过程引导学习进程，把对理论知识的学习穿插到项目的教学任务当中，项目引导、任务驱动，形式生动活泼，栏目丰富多样，并附有配套学习视频、源代码、习题、课件等丰富的教学资源。

本书既可作为计算机、软件技术相关专业移动应用开发相关课程的教材，也可作为 Android 初学者的入门教材及从事 Android 开发工程技术人员的学习参考用书。

版权专有　侵权必究

图书在版编目(CIP)数据

Android 移动应用开发项目实战教程 / 路景，赵志成主编. -- 北京：北京理工大学出版社，2022.8
ISBN 978-7-5763-0639-2

Ⅰ. ①A… Ⅱ. ①路… ②赵… Ⅲ. ①移动终端-应用程序-程序设计-教材 Ⅳ. ①TN929.53

中国版本图书馆 CIP 数据核字(2021)第 219702 号

出版发行 /	北京理工大学出版社有限责任公司
社　　址 /	北京市海淀区中关村南大街 5 号
邮　　编 /	100081
电　　话 /	(010) 68914775（总编室）
	(010) 82562903（教材售后服务热线）
	(010) 68944723（其他图书服务热线）
网　　址 /	http：//www.bitpress.com.cn
经　　销 /	全国各地新华书店
印　　刷 /	涿州市新华印刷有限公司
开　　本 /	787 毫米 × 1092 毫米　1/16
印　　张 /	20.75
字　　数 /	387 千字
版　　次 /	2022 年 8 月第 1 版　2022 年 8 月第 1 次印刷
定　　价 /	91.00 元

责任编辑 / 高　芳
文案编辑 / 高　芳
责任校对 / 周瑞红
责任印制 / 施胜娟

图书出现印装质量问题，请拨打售后服务热线，本社负责调换

前言

　　Android 系统是广泛应用于智能手机、平板电脑上的操作系统。随着技术的不断发展，可穿戴设备、智能设备等领域的应用也日渐深入。在多年的 Android 移动应用开发及相关技术课程的教学过程中，我们发现很多学生学习了知识点、技能点但不知道在实际项目中如何融会贯通、综合应用，有些学生对教材望而生畏，没有兴趣、没有动力看下去。这些问题引发了我们对于教学改革和教材编写形式的思考。伟大的人民教育家陶行知在《教学做合一》中写道："教学做是一件事，不是三件事。我们要在做上教，在做上学。在做上教的是先生；在做上学的是学生。"真正在教学、教材中落实教学做合一就是有效的突破之道。

　　在本教材的编写中，我们将教学改革中"以学生为中心、学习成果为导向、促进自主学习"的理念作为教材的编写思路，将"企业岗位任职要求、职业标准、工作过程或产品"作为教材的主体内容，将"以德树人、课程思政"的理念有机融合到教材中。本书以实战项目——"学习通关"App 设计与开发为载体，重构教学内容，将 Android 移动应用开发的相关知识和技术贯穿于项目的开发过程，让读者在项目中学，在学习中完成项目。书中设定软件技术专业的大二学生小白及其师兄为主角，以小白学习移动应用开发技术为场景，通过小白与师兄的交流推进项目开发进程及知识、技术的学习进程，希望能够引发读者的认同感、亲切感，在小白和师兄的带领下，有兴趣、有信心学习教材内容。教材中的每一个功能模块包含"功能需求描述""学习目标""若干子任务""学习目标达成度评价"4 个部分，每个模块中的子任务又按照"任务描述""任务学习目标""技术储备""任务实施""任务反思""任务总结及巩固"的环节推进，让读者知道要做什么、用什么来做、怎么做、做得好不好，真正在做中学，在学中做。

　　本教材由威海职业学院教材编写委员会系统策划，由威海职业学院软件技术专业团队与山东悦朋智能科技有限公司校企联合完成，主要参与人员有路景、赵志成、翟政凯、张诚洁、高伟等。教材编写过程中得到了院系领导、同事们及多家企业的大力支持，在此，一并表示衷心的感谢。

为了便于读者学习及使用本教材，可登录在线学习平台（www.xueyinonline.com/detail/227128246）进行在线学习及资源下载。

尽管我们做出了很多创新性的探索和努力，但书中难免会有不妥之处，欢迎各界专家和读者朋友提出宝贵意见，我们将不断改进。

编 者
2021 年 10 月
于威海职业学院笃信楼

目录

项目导读 "学习通关" App 设计与开发 1
项目准备 搭建 Android 开发环境 5
    任务 1 搭建 Android 开发环境 6
    任务 2 开发第一个 Android 程序 17
    任务 3 Android 程序调试 28
功能模块 1 用户管理功能的实现 35
    任务 1 登录界面的设计与实现 36
    任务 2 注册及详细信息界面的设计与实现 67
    任务 3 界面间跳转的实现 95
    任务 4 用户信息的存储 117
功能模块 2 限时答题功能的实现 139
    任务 1 学习积分页面的设计与实现 140
    任务 2 答题功能页面的设计与实现 152
    任务 3 使用数据库存储题目信息 180
    任务 4 倒计时功能的实现 213
功能模块 3 学习模块的实现 221
    任务 1 标题栏及导航栏的设计与实现 222
    任务 2 顶部标签栏及内容列表的设计与实现 238
    任务 3 通过网络获取资讯数据 256
功能模块 4 电台模块的实现 275
    任务 1 视频及音频播放功能的实现 276
    任务 2 服务及广播的应用 300
    任务 3 项目打包 320

项目导读

"学习通关"App设计与开发

项目背景介绍

小白是一名大学软件技术专业二年级的学生，已经学完Java程序设计的他对Android移动应用开发技术颇有兴趣，摩拳擦掌准备开始大学特学一番。恰巧，他的师兄正在一家软件公司从事Android项目研发，于是，小白在师兄的带领下开始移动应用开发技术的学习之旅。

小白：师兄，听说Android需要Java基础，你看我Java也学完了，你教教我Android移动开发呗？

师兄：挺好学嘛！就冲你这股热爱学习的劲头，这个师傅我当了！在项目里学习是最有效、最有针对性的，咱们就来个项目驱动吧。用什么项目好呢？我得好好想想……

小白趁师兄苦思冥想的时候，拿出手机开始在"学习强国"App上学习。

师兄看着小白熟练地浏览答题，灵机一动说：咱们就来做个简易版"学习强国"App怎么样？应用功能和界面操作你非常熟悉，做起来会很有动力的！咱们就叫"学习通关"App！

小白：啊？真能做得跟"学习强国"App差不多？

师兄：这就得看你用不用心了！

小白：那咱们就赶紧开始吧！我先从哪入手呢？

师兄：你别着急，咱们先从功能分析入手。

小白：这功能我已经很熟了，别浪费工夫啦！

师兄：这你就有所不知了，项目开发可不光是写代码，如果你都不知道要完成的功能是什么，代码从何写起呢？写程序只是项目开发中的一个阶段的工作，在写程序之前我们必须搞清楚要做的是什么，这就是项目需求分析。一个实际项目仅做需求分析可能就要花费两个月甚至更长的时间。虽然"学习强国"App你用得很熟练，但是现在你要研究看看这个App有哪些功能模块。明天再来找我吧。

小白一边继续操作着"学习强国"App，随口应了一声"好嘞"，一边走出门去……

想一想：请查阅资料，了解一下可以采用哪些方法进行需求分析？仔细观察"学习强国"App 包含哪些功能？试着画画功能模块图。

第二天，小白拿着画好的"学习强国"App 功能模块图来找师兄。一进门就兴高采烈地说：平时用"学习强国"App 用得挺熟练，仔细分析功能，还收获不少呢。

师兄笑着说：看来这两天功课做得不错呀！"学习强国"App 有哪些功能？说来听听。

小白从包里掏出一张纸，放在师兄面前，说：你看，这是我画的功能模块图，"学习强国"主要提供了时政要闻、科技发展、民生经济、教育文化、旅游体育等丰富多样的信息的发布、展示功能，这些信息有文字的形式、有视频和音频的形式，还有学习答题的功能以及对用户登录信息进行管理的功能。

师兄说：不错，你基本总结出了"学习强国"App 的大部分功能，这是一个非常丰富的学习平台。这么多功能我们不可能一一实现，我们就从中选择一些能够包含 Android 开发基本技术的模块来完成我们自己的"学习通关"App 吧。

功能模块 1：用户信息管理功能

功能模块2：限时答题功能

功能模块3：学习模块

功能模块4：电台模块

小白兴奋地说：师兄，我们快开始吧。

师兄：别急，我们先来看看每个功能模块的任务分解和用到的技术，让你心中有数。

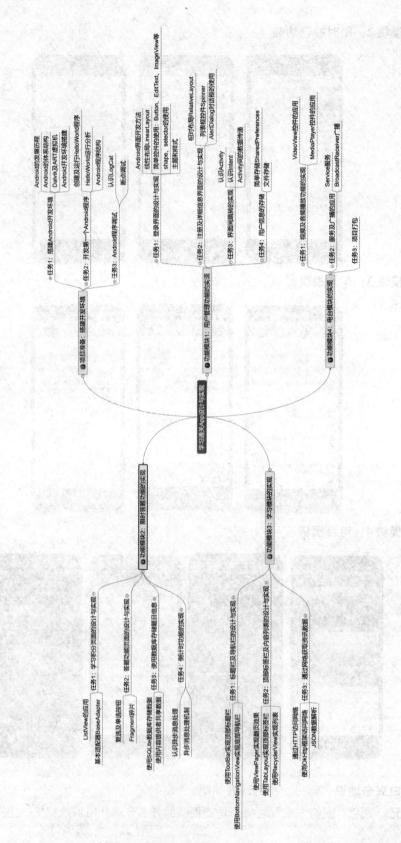

项目准备

搭建Android开发环境

> **说在前面**
>
> **师兄**：在进行任何一项开发之前，我们都需要对所用到的技术进行了解，搭建开发环境并熟悉开发环境的使用，这就是俗话说的"工欲善其事，必先利其器"。
>
> **小白**：我现在的手机就是 Android 系统，对 Android 系统的使用我可是非常熟悉的，而且我学习过 Java 开发，对 Eclipse 的使用也很熟悉了。
>
> **师兄**：只会用 Android 系统可不够，还需要对 Android 系统有更深的认识；Eclipse 虽然也可以进行 Android 开发，但谷歌（Google）公司已经终止了对 Eclipse 的支持。Android Studio 目前才是主流的开发环境。咱还是别着急，一步一个脚印，按照下面 3 个任务踏实前行吧。
>
> 任务 1：搭建 Android 开发环境。
>
> 任务 2：开发第一个 Android 程序。
>
> 任务 3：Android 程序调试。

学习目标

1. 知识目标

（1）了解 Android 的发展历史。

（2）了解 Android 的体系结构。

（3）掌握 Android 程序结构。

2. 技能目标

（1）能够搭建 Android Studio 开发环境。

（2）能够创建并运行 Android 程序。

（3）能够对 Android 程序进行调试。

任务 1　搭建 Android 开发环境

任务描述

根据师兄的介绍，了解 Android 系统的发展历程及体系结构，完成主流 Android 开发环境的搭建及配置，为后期开发做好准备。

任务学习目标

通过本任务需达到以下目标：
- 了解 Android 的发展历程及主流版本。
- 了解 Android 的体系结构。
- 能够搭建 Android 开发环境并解决环境搭建过程中遇到的各种问题。

技术储备

认识 Android

一、Android 的发展历程

众所周知，Android 系统是谷歌（Google）公司的一款手机操作系统，但 Android 并不是由谷歌研发出来的，安迪·鲁宾（Andy Rubin）成立了 Android（安卓）企业，并带领他的团队打造了 Android 操作系统。Google 公司在 2005 年收购了这个仅成立 22 个月的企业，并让安迪·鲁宾成为 Google 公司的工程部副总裁，继续负责 Android 项目的研发工作。

2007 年 11 月，Google 公司对外界展示了这款名为"Android"的操作系统，并与 84 家硬件制造商、软件开发商及电信运营商组建了开放手机联盟，共同研发、改良 Android 系统。随后，Google 公司以 Apache 开源许可证的授权方式，发布了 Android 的源代码。

Android 版本升级非常快，几乎每隔半年就会发布一个新的版本。2021 年 5 月，最新的版本 Android 12 发布。有趣的是，在 Android 10 之前，每一个 Android 版本都会有一个以甜品命名的代号，但由于并不是每种甜品都能与全部用户产生共鸣，从 Android 10 之后，Google 公司改变了这个传统。Android 各版本发布时间及其代号如图 0 - 1 所示。

> **长知识：**
>
> **Android 图标的由来**
>
> Android 一词最早出现于法国作家利尔亚当（Auguste Villiers de l'Isle Adam）在 1886 年发表的科幻小说《未来夏娃》中，将外表像人的机器起名为"Android"。Android 本意指"机器人"，Google 公司将 Android 的标识设计为一个绿色机器人，表示 Android 系统符合环保概念。

二、Android 的体系结构

Android 系统采用分层架构，由高到低分为 4 层，依次是应用程序层（Application）、应用程序框架层（Application Framework）、核心类库（Libraries）和 Linux 内核（Linux Kernel），如图 0 - 2 所示。

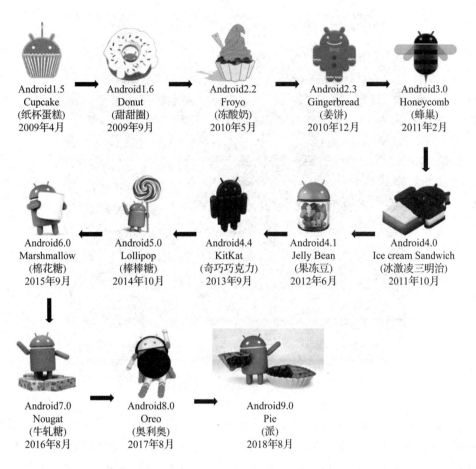

图 0-1 Android 各版本及代号

每层的具体功能如下。

(一) 应用程序层 (Application)

应用程序层是一个核心应用程序的集合,所有安装在手机上的应用程序都属于这一层,例如系统自带的联系人程序、短信程序,或者从 Google Play 上下载的小游戏等都属于应用程序层。

(二) 应用程序框架层 (Application Framework)

应用程序框架层主要提供了构建应用程序时用到的各种 API (Application Programming Interface,应用程序接口)。Android 自带的一些核心应用就是使用这些 API 完成的,例如活动管理器 (Activity Manager)、通知管理器 (Notification Manager)、内容提供者 (Content Provider) 等,开发者也可以通过这些 API 开发应用程序。

(三) 核心类库 (Libraries)

核心类库中包含了系统库及 Android 运行环境 (Android Runtime)。

(1) 系统库主要是通过 C/C++ 库来为 Android 系统提供主要的特性支持,如 OpenGL/EL 库提供了 3D 绘图的支持,WebKit 库提供了浏览器内核的支持。

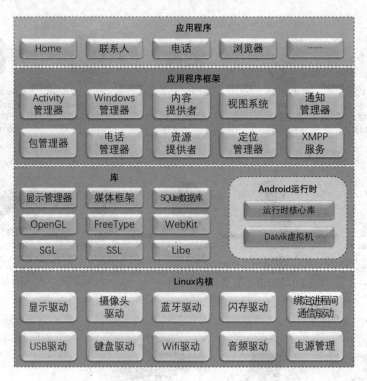

图 0-2 Android 体系结构

（2）Android 运行时库主要提供了一些核心库，能够允许开发者使用 Java 语言来编写 Android 应用程序，另外，Android 运行时库中还包括了 Dalvik 虚拟机，它使得每一个 Android 应用程序都能运行在独立的进程当中，并且拥有一个自己的 Dalvik 虚拟机实例，相较于 Java 虚拟机，Dalvik 是专门为移动设备定制的，它针对手机内存、CPU 性能等做了优化处理。

（四）Linux 内核（Linux Kernel）

Linux 内核为 Android 设备的各种硬件提供了底层的驱动，如显示驱动、音频驱动、照相机驱动、蓝牙驱动、电源管理驱动等。

三、Dalvik 虚拟机

Android 应用程序的主要开发语言是 Java，它通过 Dalvik 虚拟机来运行 Java 程序。Dalvik 是 Google 公司设计的用于 Android 平台的虚拟机，其指令集基于寄存器架构，执行其特有的 dex 文件来完成对象生命周期管理、堆栈管理、线程管理、安全异常管理、垃圾回收等重要功能。

每一个 Android 应用程序在底层都会对应一个独立的 Dalvik 虚拟机实例，其代码在虚拟机的解释下得以执行。Dalvik 虚拟机与 Java 虚拟机有很多不同之处，具体如图 0-3 所示。

从图 0-3 中可以看出，JVM 是基于栈的，Java 程序经过编译生成字节码文件，JVM 通过解释执行程序；Dalvik 虚拟机运行的是 dex 字节码文件，这种文件是 Java 源文件经过 JDK

编译器编译成 class 文件之后，通过 dx 工具将部分 class 文件打包生成的。dex 格式是专为 Dalvik 设计的一种压缩格式，适用于内存和处理器资源有限的系统。

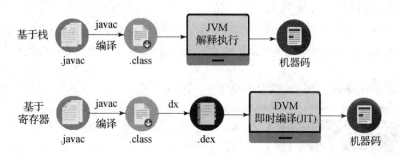

图 0-3 JVM 虚拟机与 Dalvik 虚拟机对比

Dalvik 虚拟机采用即时编译将 dex 字节码转换为机器码，即每次运行的时候都要进行编译运行，这种机制使运行的效率降低，但应用安装比较快，而且更容易在不同的硬件和架构上运行。

Android 程序都运行在一个 Dalvik 虚拟机实例中，而每一个 Dalvik 虚拟机实例都是一个独立的进程空间，每个进程之间可以通信。

四、ART 虚拟机

为了提升 Android 系统下应用程序运行的流畅性，Android 4.4 引入了 ART 虚拟机，并在 Android 5.0 版本中默认启用 ART。

ART 模式与 Dalvik 模式最大的不同在于，在启用 ART 模式后，系统在安装应用程序时会进行一次预编译（Ahead Of Time compilation，AOT），将字节码预先编译成机器码并存储在本地，这样程序在每次运行时就不需要执行编译了，运行效率会大大提升，设备耗电也会降低。但采用预编译机制也有两个主要缺点：一是应用程序安装时间变长，尤其是一些复杂的应用；二是字节码预先编译成机器码，机器码需要的存储空间也更大。

> 长知识：
>
> **ART 虚拟机的改进**
>
> Android 7.0 中，ART 虚拟机也加入了即时编译器 JIT，作为预编译 AOT 的补充，在安装应用程序时并不会将字节码全部编译成机器码，而是在运行中将热点代码编译成机器码，从而缩短应用的安装时间并节省了存储空间。

任务实施

如师兄所言，最开始 Android 使用 Eclipse 作为开发工具，2015 年年底，Google 公司声明不再对 Eclipse 提供支持服务，后起之秀 Android Studio 逐渐取代 Eclipse。现在开始搭建 Android 开发环境，步骤如下。

1. 检查 Android Studio 安装环境

（1）操作系统为 Windows 7 或更高版本。

（2）JDK 版本不低于 1.7。

（3）系统空闲内存不少于 2 GB。

2. 下载 Android Studio 安装包

Android Studio 安装包可以从中文社区下载，我们可以根据系统平台下载最新版本的安装包。64 位 Windows 系统单击框中标识的位置进行下载即可，如图 0-4 所示。

图 0-4　下载页面

3. 安装 Android Studio

（1）双击安装包的 exe 文件，进入"Android Studio Setup"对话框，如图 0-5 所示。

图 0-5　"Android Studio Setup"对话框

（2）单击"Next"按钮，进入"Choose Components"对话框，如图 0-6 所示。

（3）单击"Next"按钮，进入"Configuration Settings"对话框，如图 0-7 所示。图中的输入框用于设置 Android Studio 的安装路径，单击"Browse"按钮可更改安装路径。这里我们选择不更改路径，使用系统默认设置路径。

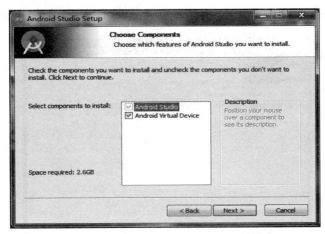

图 0-6 "Choose Components" 对话框

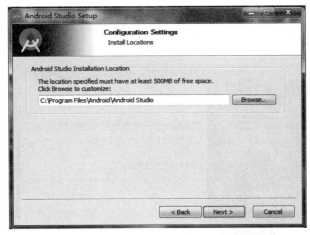

图 0-7 "Configuration Settings" 对话框

(4) 单击 "Next" 按钮，进入 "Choose Start Menu Folder" 对话框，设置在 "开始" 菜单中显示的文件夹名称，如图 0-8 所示。

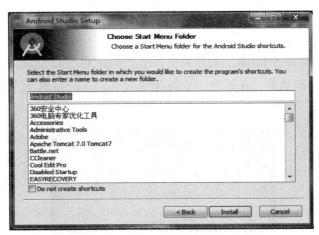

图 0-8 "Choose Start Menu Folder" 对话框

(5) 单击"Install"按钮，进入"Installing"对话框开始安装，如图 0 – 9 所示。

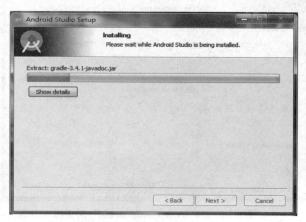

图 0 – 9　"Installing"对话框

(6) 安装完成后，单击"Next"按钮，进入"Completing Android Studio Setup"对话框，如图 0 – 10 所示。

图 0 – 10　"Completing Android Studio Setup"对话框

(7) 单击"Finish"按钮，Android Studio 安装全部完成。

4. 配置 Android Studio

(1) 如果我们在单击"Finish"按钮时勾选了"Start Android Studio"选项，安装完成之后Android Studio 会自动启动，并弹出一个选择导入 Android Studio 配置文件夹位置的对话框，如图 0 – 11 所示。

图 0 – 11　选择导入 Android Studio 配置文件夹位置

图 0-11 中包含两个选项，第 1 个选项表示自定义 Android Studio 配置文件夹的位置，第 2 个选项表示不导入配置文件夹的位置。如果之前安装过 Android Studio，想要导入之前的配置文件夹的位置，可以选择第 1 项，否则，选择第 2 项。这里我们选择第 2 项。

（2）单击"OK"按钮，进入 Android Studio 开启窗口，如图 0-12 所示。

图 0-12 Android Studio 开启窗口

进度完成之后，会弹出"Android Studio First Run"对话框，如图 0-13 所示。

这是因为是第一次安装 Android Studio，启动后检测到默认安装的文件夹中没有 SDK，如果单击"Setup Proxy"按钮，会立即在线下载 SDK。单击"Cancel"按钮，则可以后续使用时再下载 SDK。这里我们单击"Cancel"按钮进入"Android Studio Setup Wizard"对话框，如图 0-14 所示。

图 0-13 "Android Studio First Run"对话框

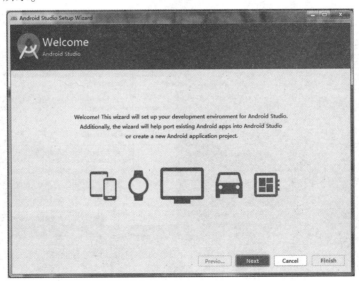

图 0-14 "Android Studio Setup Wizard"对话框

(3) 单击"Next"按钮，进入"Install Type"对话框，如图 0-15 所示。

图 0-15 中包含"Standard"和"Custom"两个选项，分别表示安装 Android Studio 的标准设置和自定义设置。这里推荐选择"Standard"选项，默认安装好开发 Android 程序的基本开发配置。

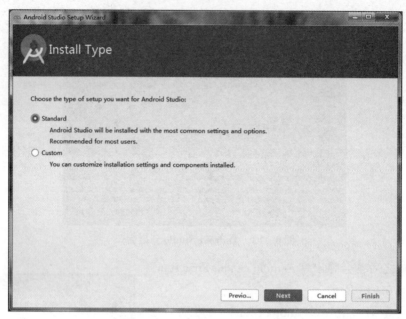

图 0-15 "Install Type"对话框

(4) 单击"Next"按钮，进入"Select UI Theme"对话框，如图 0-16 所示。"Darcula"和"Intellij"是黑白两种主题颜色，可以根据自己的喜好自行选择。

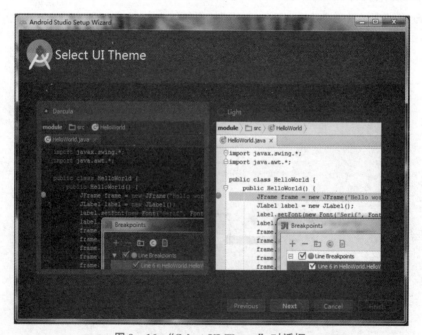

图 0-16 "Select UI Theme"对话框

（5）单击"Next"按钮，进入"Verify Settings"对话框，如图 0-17 所示。

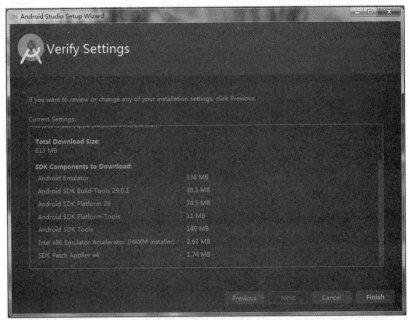

图 0-17 "Verify Settings"对话框

图中可以看到需要下载的 SDK 组件。此时如果想查看或更改前面的安装设置，单击"Previous"按钮即可，如果不想下载窗口中显示的 SDK 组件，则单击"Cancel"按钮即可，否则单击"Finish"按钮下载 SDK 组件。

（6）单击"Finish"按钮，进入"Downloading Components"对话框，如图 0-18 所示。

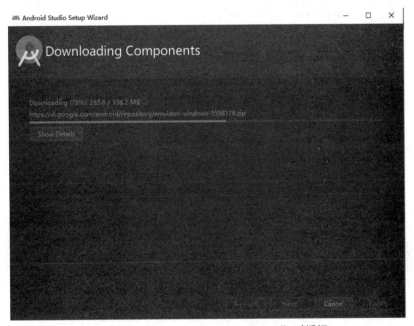

图 0-18 "Downloading Components"对话框

下载完成后，会显示下载完成窗口。

（7）单击"Finish"按钮，进入"Welcome to Android Studio"对话框，如图 0 – 19 所示。

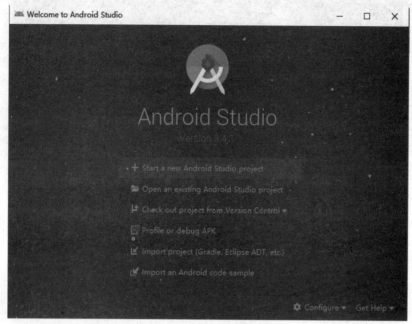

图 0 – 19　"Welcome to Android Studio"对话框

至此，Android Studio 工具的配置全部完成。

任务反思

按照上述步骤安装并配置 Android Studio 开发环境，如果安装及配置过程顺利，并出现 Welcome to Android Studio 窗口，则安装配置成功。如果在安装及配置过程中出现问题，请将问题记录在下方，并通过网络查找问题原因及解决方案。

问题 1：＿＿＿＿＿＿＿＿＿＿＿＿＿＿＿＿＿＿＿＿＿＿＿＿＿＿＿＿＿＿＿＿＿＿

产生原因：＿＿＿＿＿＿＿＿＿＿＿＿＿＿＿＿＿＿＿＿＿＿＿＿＿＿＿＿＿＿＿＿

＿＿＿＿＿＿＿＿＿＿＿＿＿＿＿＿＿＿＿＿＿＿＿＿＿＿＿＿＿＿＿＿＿＿＿＿＿＿

解决方法：＿＿＿＿＿＿＿＿＿＿＿＿＿＿＿＿＿＿＿＿＿＿＿＿＿＿＿＿＿＿＿＿

＿＿＿＿＿＿＿＿＿＿＿＿＿＿＿＿＿＿＿＿＿＿＿＿＿＿＿＿＿＿＿＿＿＿＿＿＿＿

问题 2：＿＿＿＿＿＿＿＿＿＿＿＿＿＿＿＿＿＿＿＿＿＿＿＿＿＿＿＿＿＿＿＿＿＿

产生原因：＿＿＿＿＿＿＿＿＿＿＿＿＿＿＿＿＿＿＿＿＿＿＿＿＿＿＿＿＿＿＿＿

＿＿＿＿＿＿＿＿＿＿＿＿＿＿＿＿＿＿＿＿＿＿＿＿＿＿＿＿＿＿＿＿＿＿＿＿＿＿

解决方法：＿＿＿＿＿＿＿＿＿＿＿＿＿＿＿＿＿＿＿＿＿＿＿＿＿＿＿＿＿＿＿＿

任务总结及巩固

师兄：怎么样？Android 的 "前世今生" 都了解了吧！

小白：嗯！作为一个移动操作系统，Android 的发展真是非常迅速。但是我印象最深的还是它每个版本的代号，虽然都说写程序是个枯燥无味的工作，但是这些可爱的代号还真是让我看到了程序员的另一面。

师兄：你可别觉得程序员都是满脑子代码、无趣的人，很多程序员都很有才华。话说回来，在环境搭建过程中，你也要注意我们可能遇到的各种问题，要善于查找资料、多总结，很多问题就会迎刃而解了。我给你出几道小题目，看看你掌握得怎么样。

1. Dalvik 中的 dx 工具会把部分 class 文件转换成_____文件。
2. Dalvik 虚拟机是基于_____的架构。
3. Dalvik 虚拟机属于 Android 系统架构中的_____层。
4. Android 中的短信、联系人管理、浏览器等属于 Android 系统架构中的_____层。
5. Dalvik 是 Google 公司设计的用于 Android 平台的虚拟机。(　　)
6. Android 应用程序的主要开发语言是 Java 和 Kotlin。(　　)
7. Android 系统采用分层架构，分别是应用程序层、应用程序框架层、核心类库和 Linux 内核。(　　)

任务 2　开发第一个 Android 程序

任务描述

创建第一个 Android 程序，并通过模拟器运行该程序。

任务学习目标

通过本任务需达到以下目标：
➢ 能够创建 Android 应用程序。
➢ 能够创建模拟器并运行程序。
➢ 能够下载并更新 SDK。
➢ 理解 Android 程序结构及各部分的作用。

任务实施

1. 创建 HelloWorld 程序

（1）单击 "Welcome to Android Studio" 对话框中的 "Start a new Android Studio project" 选项，进入 "Create New Project" 对话框，如图 0 - 20 所示。

此页面包含 5 个选项标签，分别代表 5 种项目类别：手机和平板项目（Phone and Tablet）、可穿戴设备项目（Wear OS）、TV 项目（TV）、针对汽车开发的项目（Android Auto）及针对物联网设备的项目（Android Things）。

如何创建 Android 项目

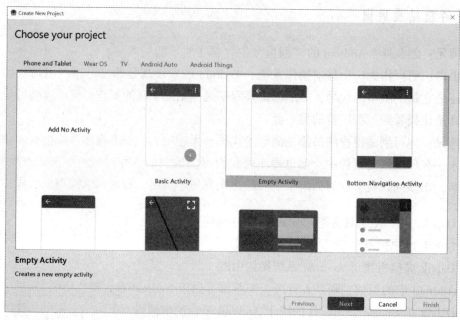

图 0-20 "Create New Project" 对话框

这里我们选择第一个项目类别下的"Empty Activity"。该类型的 Activity 界面上没有放任何控件。其他类型的 Activity 都是在 Empty Activity 类型的基础上添加了其他功能,可以根据实际需求使用不同类型的 Activity。

(2) 单击"Next"按钮,进入"Configure your project"对话框,如图 0-21 所示。

此窗口中我们设置项目名为"HelloWorld",包名为"com.example.helloworld",存储位置为"D:\AndroidWorkspace\DemoSpace\HelloWorld",编程语言为"Java",最小 API 版本为"API 19:Android 4.4 (kitkat)"。

长知识:

Kotlin 语言

Kotlin 语言是 Android 官方支持的开发语言,它是一个用于现代多平台应用的静态编程语言,由 Java IDEA IntelliJ 的提供商 JetBrains 开发。Kotlin 程序可以编译成 Java 字节码,也可以编译成 JavaScript,方便在没有 JVM 的设备上运行。

(3) 单击"Finish"按钮,进入 Android Studio 工具的编辑窗口,如图 0-22 所示。

当前窗口中间为界面设计视图,单击下方的视图切换标签,可将设计视图切换为代码视图,如图 0-23 所示。

至此,HelloWorld 程序的创建已全部完成。

2. 创建 Android 模拟器

Android 程序可以运行在手机、平板电脑等真实物理设备上,在开发过程中如果没有真实物理设备可以使用 Android 模拟器替代。模拟器是一个运行在计算机上的虚拟设备,通过模拟器可以预览和测试 Android 应用程序。

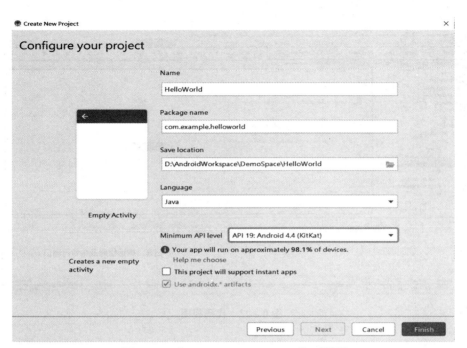

图 0-21 "Configure your project"对话框

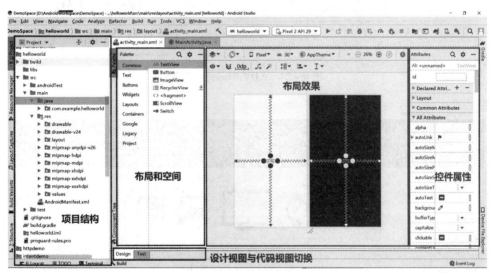

图 0-22 编辑窗口

(1) 单击图 0-24 所示的 ADV Manager 图标，进入"Your Virtual Devices"对话框，如图 0-25 所示。

(2) 单击"Create Virtual Device"按钮，进入"Select Hardware"对话框，如图 0-26 所示。

如何运行项目

图 0-26 中，左侧 Category 为设备类型，中间为设备的名称、尺寸大小、分辨率、密度等信息，右侧是设备的预览图。这里我们选择"Phone""Pixel 2"。

Android 移动应用开发项目实战教程

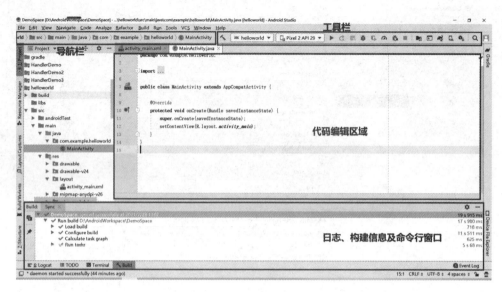

图 0-23 代码视图

图 0-24 ADV Manager 图标

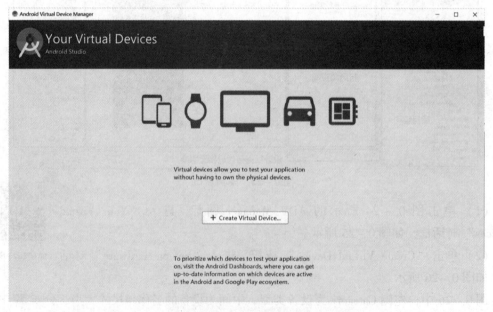

图 0-25 "Your Virtual Devices" 对话框

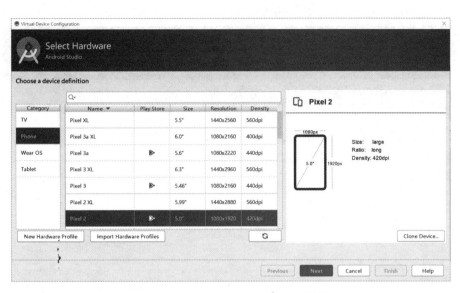

图 0-26 "Select Hardware" 对话框

(3) 单击"Next"按钮,进入选择设备的"System Image"对话框,如图 0-27 所示。

图 0-27 "System Image" 对话框

图 0-27 中,左侧为推荐的 Android 系统镜像,右侧为选中的 Android 系统镜像对应的图标。这里我们选择 Android 10 的系统版本进行下载。

(4) 单击版本名称后的"Download"按钮,进入"License Agreement"对话框。

(5) 选中"Accept"选项接受窗口中显示的信息,单击"Next"按钮进入"Component Installer"对话框,如图 0-28 所示。

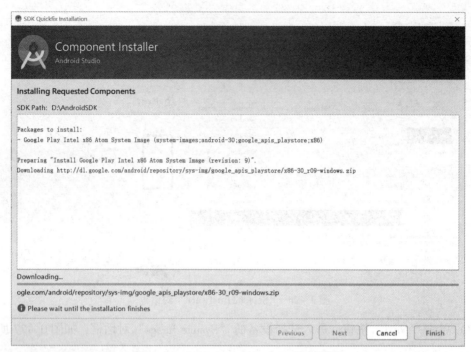

图 0-28 "Component Installer" 对话框

（6）单击"Finish"按钮关闭当前窗口并返回"System Image"对话框，选中系统版本名称为 Q 的条目，单击"Next"按钮进入"Android Virtual Device（AVD）"对话框，如图 0-29 所示。

图 0-29 "Android Virtual Device（AVD）"对话框

（7）单击"Finish"按钮完成模拟器的创建，如图 0-30 所示。
（8）单击模拟器后的"运行"按钮，启动模拟器。

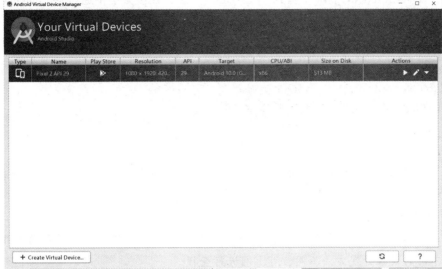

图 0 – 30　模拟器创建成功

3. 运行 HelloWorld 应用程序

单击图 0 – 31 工具栏中的"运行"按钮，应用程序就会运行在模拟器上。

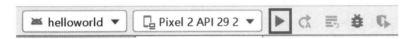

图 0 – 31　运行程序

4. HelloWorld 运行分析

前面我们创建并运行了 HelloWorld 应用程序，却一行代码都没有写，是如何运行出效果的呢？下面我们对该应用程序进行分析。

当 Android 程序运行时，系统首先会查找配置文件 AndroidManifest.xml，该文件中注册了所有应用程序中用于显示的 Activity（可以理解为界面），如果某个 Activity 节点中＜Intent – filter＞中有如下两行代码：

```
< action android:name = "android.intent.action.MAIN" / >
< category android: name = " android.intent.category.LAUNCHER" / >
```

那么程序运行时，这个 Activity 将会作为程序的第一个页面显示。在 HelloWorld 中，MainActivity 就是这样的页面。因此，第一个页面锁定了 MainActivity。

```
< activity android:name = ".MainActivity" >
< intent – filter >
    < action android:name = "android.intent.action.MAIN" / >
    < category android:name = "android.intent.category.LAUNCHER" / >
</intent – filter >
</activity >
```

接着，在 MainActivity.java 中找到 OnCreate( )方法，在该方法中通过 setContentView( )方法加载了 res/layout 目录中的 activity_main.xml 布局文件，从而在运行时对布局进行解析，产生了我们看到的运行效果。

```
protected void onCreate(Bundle savedInstanceState) {
    super.onCreate(savedInstanceState);
    setContentView(R.layout.activity_main);
}
```

长知识：

除了使用 Android Studio 中自带的模拟器，还可以使用手机真机及第三方模拟器进行程序运行。

1. 使用手机真机

通过数据线将手机与计算机连接，进入手机的开发者选项，开启设备的 USB 调试模式。不同的设备开发方法不同，可上网进行查找。

2. 使用第三方模拟器

在计算机上安装夜神、MuMu 等第三方模拟器。开启模拟器，在设备列表中即可找到第三方模拟器。有时会出现第三方模拟器无法连接的问题，可以通过以下方法解决。

（1）配置环境变量。

将 SDK 目录下 platform – tools 路径配置到系统环境变量 Path 中。

（2）输入连接命令。

在 Android Studio 下方的 Terminal 终端，输入连接命令"adb connect 127.0.0.1：端口号"，图 0 – 32 中"62001"是连接夜神模拟器的端口号，不同的模拟器，命令最后的端口号不同，可自行上网查找。

图 0 – 32　输入连接命令

注意： SDK 目录在哪里找？

在安装 Android Studio 时已经附带安装了 SDK，Android SDK 会不断进行更新，如果想要安装最新版本或者之前版本的 SDK，则需要重新下载相应版本的 SDK。最简单的方法就是在 Android Studio 中下载，在这里也能找到 SDK 的安装目录。具体做法如下。

（1）单击如图 0-33 所示的导航栏中的 SDK Manager 图标，进入"Default Settings"对话框。

图 0-33　SDK Manager 图标

（2）左侧选择 Android SDK，右侧为 Android SDK 可设置的一些选项，如图 0-34 所示。

- Android SDK Location：Android SDK 的安装路径。
- SDK Platforms：Android SDK 的版本信息。
- SDK Tools：Android SDK 的工具集合，包括构建工具、模拟器镜像等。

可以勾选需要下载的 SDK 版本和 Tools 工具，单击"OK"按钮下载即可。

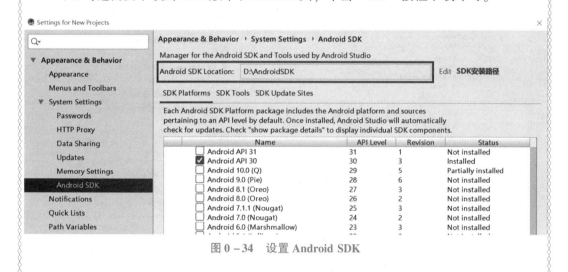

图 0-34　设置 Android SDK

技术储备

Android 程序结构

Android Studio 为我们提供了多种目录结构，其中我们最常用的就是 Project 和 Android 两种，下面我们具体来看一下两种目录结构下的应用程序具体组成结构。

（一）Project 结构

Project 结构下的目录结构如图 0-35 所示。app 目录用于存放程序的代码和资源等内容，它包含很多子目录，具体如下。

- libs：存放第三方 jar 包。
- src/androidTest：存放测试的代码文件。
- src/main/java：存放程序的源代码文件。
- src/main/res：存放程序的资源文件。

- src/AndroidManifest.xml：它是整个程序的配置文件，在该文件中配置程序所需权限和注册程序中用到的四大组件。
- app/build.gradle：该文件是 App 的 gradle 构建脚本。在该文件中有 3 个重要属性，分别为 compileSdkVersion、minSdkVersion、targetSdkVersion，分别表示编译的 SDK 版本、支持的最低版本、支持的目标版本。
- build.gradle：该文件是程序的 gradle 构建脚本。
- local.properties：该文件用于指定项目中所使用的 SDK 路径。

（二）Android 结构

Android 项目目录结构

Android 结构下的目录结构如图 0-36 所示。其目录结构相对 Project 结构简单明了许多，具体目录内容及文件含义与 Project 下的相应目录一致，这里就不再赘述。

图 0-35　Project 目录结构

图 0-36　Android 目录结构

> **长知识：**
>
> **区分 Android Studio 中的 Project 和 Module**
>
> 我们刚才创建 HelloWorld 应用程序时创建了一个 Project，是一个用来存放 Android 项目的文件夹。Module 则代表一个独立的 Android 项目或一个模块，在 HelloWorld 中的 App 就是一个在 Module 中的 Android 项目。
>
> 我们可以在一个 Project 中通过依次选择 File→New→New Module 创建一个新的 Module，使得一个 Project 中包含多个 Module。这些 Module 可以是互相独立的不同的 Android 项目，也可以是一个 Module 作为一个模块被其他多个 App 公用。

项目准备　搭建 Android 开发环境

任务反思

按照上述步骤创建并运行第一个 Android 应用程序，如果在创建及运行中出现问题，请将问题记录在下方，并通过网络查找问题原因及解决方案。

```
问题 1：_____
产生原因：_____
_____
解决方法：_____
_____

问题 2：_____
产生原因：_____
_____
解决方法：_____
_____
```

任务总结及巩固

师兄：小白，你的第一个 Android 应用程序就运行起来了。虽然过程并不难，但是你需要重点关注以下几方面。

- 熟练掌握项目创建过程中每个步骤的含义及选择项。
- 能够配置 Android 模拟器并让程序运行起来。
- 清楚 Android 项目的目录结构及每种文件及资源所在的位置。

小白：嗯！Android 程序比我以前写的 Java 程序都神奇。我一行代码都还没写，HelloWorld 竟然就能运行了。Android 的项目结构也比以前的 Java 程序复杂很多，我还有点分不太清楚呢。

师兄：不积跬步，无以至千里，慢慢来，不着急，多练多看自然就熟悉了。

1. Android 项目中的 res 目录用来存放（　　）。
   A. Java 源文件　　　　　　　　B. 资源文件
   C. 配置文件　　　　　　　　　D. 测试代码

2. 下列方法中，（　　）用来加载布局文件。
   A. setContentView( )　　　　　B. onCreate( )
   C. setLayout( )　　　　　　　　D. onResume( )

3. R. layout. activity_main 代表（　　）。

A. 控件 ID
B. 布局文件 ID
C. 图片 ID
D. 颜色资源 ID

任务 3　Android 程序调试

任务描述

通过 Logcat 工具对 Android 应用程序进行调试，并在程序中设置断点，通过断点调试的方式查看程序执行过程，发现程序运行中存在的问题。

任务学习目标

通过本任务需达到以下目标：
➢ 能够通过 Logcat 对 Android 程序进行调试。
➢ 能够通过设置断点，对程序运行过程进行监控。

技术储备

一、认识 Logcat

Logcat 是 Android 中的日志控制台，用于获取程序从启动到关闭的日志信息，Android 程序在设备中运行时，程序的调试信息就会输出到该设备单独的日志缓冲区中，要想从设备日志缓冲区中获取信息，就需要学会使用 Logcat。

Android 使用 android.util.Log 类的静态方法实现输出程序的调试信息，Log 类所输出的日志内容有 6 个级别，由低到高分别是 Verbose（全部信息）、Debug（调试信息）、Info（一般信息）、Warning（警告信息）、Error（错误信息）、Assert（显示断言失败后的错误信息），这些级别分别对应 Log 类中的 Log.v( )、Log.d( )、Log.i( )、Log.w( )、Log.e( )、Log.wtf( ) 静态方法。可以在 Android 应用程序中通过上述静态方法输出调试中需要使用的信息。

同时，程序运行过程中出现的错误及异常信息也会在 Logcat 中显示出来，因此，Logcat 是我们在调试程序时的重要工具。

二、断点调试

我们可以在程序中必要的位置设置断点，采用 Debug 模式运行程序，让程序中断在设置断点的地方。通过单步调试，让程序逐句执行，观察程序的执行顺序、变量的变化情况，对程序运行过程进行分析。相比逐句或逐条指令地检查代码，断点调试可以更为准确、细致地观察程序的执行过程，大大加快程序调试过程。这是程序员的必备技能之一。

任务实施

1. 利用 Logcat 输出并观察日志信息

（1）在 HelloWorld 的 MainActivity 中加入日志信息输出代码。

```
1  public class MainActivity extends AppCompatActivity {
2      @Override
3      protected void onCreate(Bundle savedInstanceState) {
4          super.onCreate(savedInstanceState);
5          setContentView(R.layout.activity_main);
6          Log.v("MainActivity","Verbose");
7          Log.d("MainActivity","Debug");
8          Log.i("MainActivity","Info");
9          Log.w("MainActivity","Warning");
10         Log.e("MainActivity","Error");
11         Log.wtf("MainActivity","Assert");
12     }
13 }
```

（2）运行程序，观察下方 Logcat 窗口中的 Log 信息，如图 0 – 37 所示。

图 0 – 37　Log 信息

（3）设置 Logcat 过滤器。

单击右侧下拉按钮，选择"Edit Filter Configuration"选项，如图 0 – 38 所示。

图 0 – 38　选择"Edit Filter Configuration"选项

进入"Create New Logcat Filter"对话框，如图 0 – 39 所示。

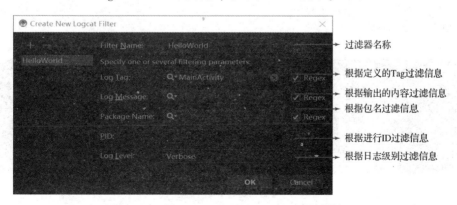

图 0 – 39　"Create New Logcat Filter"对话框

设置过滤器后的 Logcat 信息如图 0-40 所示。

图 0-40　设置过滤器后的 Logcat 信息

2. 对程序进行断点调试

（1）在 MainActivity 中加入一段 for 循环程序。

```
1   public class MainActivity extends AppCompatActivity {
2       @Override
3       protected void onCreate(Bundle savedInstanceState) {
4           super.onCreate(savedInstanceState);
5           setContentView(R.layout.activity_main);
6           for(int i =0;i <10;i + +) {
7               //获取当前 i 的值
8               int selector = i;
9               //打开 Log 查看当前 i 的值
10              Log.i("断点调试","for 当前的 i 的值:" + i);
11              //调用方法
12              stepNext(i);
13          }
14      }
15      public void stepNext(int i){
16          Log.i("断点调试","stepNext 当前的 i 的值:" + i);
17      }
18  }
```

（2）设置断点。在"int selector = i"语句行号后的空白处单击，出现红点即完成断点的设置，如图 0-41 所示。

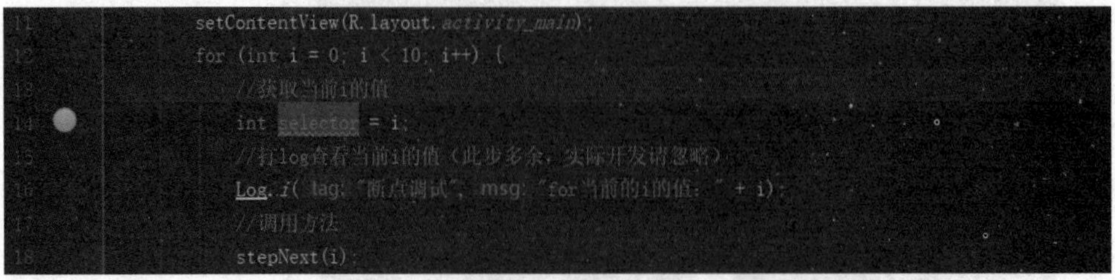

图 0-41　设置断点

（3）Debug 模式运行。单击工具栏上的如图 0 - 42 所示的 Debug 图标，进入调试模式，并查看下方的调试面板，如图 0 - 43 所示。

图 0 - 42　Debug 图标

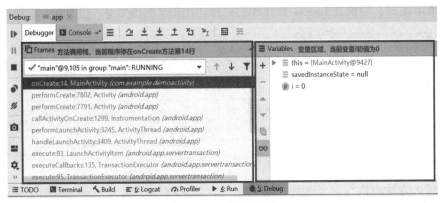

图 0 - 43　调试面板

在当前程序运行位置，i 的值已经在代码中展现出来。此时，设置断点的第 14 行还未执行，如图 0 - 44 所示。

图 0 - 44　当前程序运行位置

（4）单步调试。单击如图 0 - 45 所示的 step over（单步调试）图标或按【F8】键，逐条程序往下运行。

图 0 - 45　step over 图标

第 14 行执行完毕，selector 变量的值为 0，变量区域中 selector 变量的值显示为 0，如图 0 - 46 所示。

图 0-46　执行程序并观察变量区域

继续单步调试，切换到 Logcat 查看日志，打印出的 i 的值是 0，如图 0-47 所示。

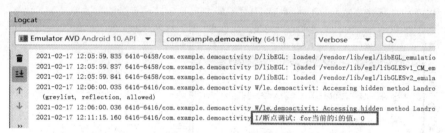

图 0-47　查看日志

(5) 转入方法内执行。当遇到方法调用时，如果继续单击 step over 图标，则执行下一条语句；如果需要转入方法内执行，观察方法的执行过程，则需单击如图 0-48 所示的 step into 图标或按【F7】键。

图 0-48　step into 图标

for 循环中调用了一个 stepNext() 方法，转入方法内执行。此时，转到 stepNext() 方法内执行，如图 0-49 所示。

图 0-49　转入方法内执行

在下方的方法调用栈中可以看到目前执行的方法为"stepNext"，右侧变量区域中 i 为 stepNext() 方法中的参数 i，其当前值为 0，如图 0-50 所示。

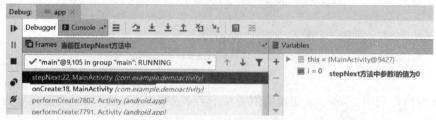

图 0-50　查看信息

任务反思

按照上述断点调试方法逐步执行程序并通过 Logcat 观察输出信息,将调试过程中的变量变化情况及输出情况记录在下方。

```
_____
_____
_____
```

任务总结及巩固

> **师兄**:程序调试是程序员的基本功,我接触过好多新手实习生,程序出错时便手忙脚乱,从来不看日志信息,更不会断点调试,一脸茫然地盯着程序到处乱找。所以,程序调试能力得从平时学习的一点一滴积累起来,最好能做到以下几点。
> - 程序出错莫慌张,翻看日志找线索。
> - 必要信息可输出,通过信息辨问题。
> - 断点调试要掌握,执行过程觅踪迹。
>
> **小白**:我以前在程序出现问题的时候就经常摸不着头脑,像没头苍蝇一样乱找。看来巧妙利用集成开发环境中的工具可以让程序调试事半功倍。
>
> **师兄**:工具是必须要会用的,但是对出错信息及执行过程的分析才是我们发现问题的关键,这种能力可不是一天就能练成的。来,给你制造个小问题,通过任务中学习的内容看你能不能发现问题在哪里。

下面的程序是否能够正常运行?如果程序出错,出错的现象是什么?Logcat 中的报错信息是什么?

```java
1  public class MainActivity extends AppCompatActivity {
2      private TextView textView;
3      @Override
4      protected void onCreate(Bundle savedInstanceState) {
5          super.onCreate(savedInstanceState);
6          setContentView(R.layout.activity_main);
7          textView.setText("你好 Android");
8      }
9  }
```

学习目标达成度评价

了解对模块的自我学习情况,完成表 0-1。

表 0-1　学习目标达成度评价表

序号	学习目标	学生自评
1	了解 Android 的发展历史	□了解 □基本了解 □不了解
2	了解 Android 的体系结构	□了解 □部分了解 □不了解
3	熟悉 Android 的程序结构及各部分的作用	□熟悉 □基本熟悉 □不熟悉
4	能够搭建 Android Studio 开发环境	□已顺利搭建环境 □已经安装但仍存在问题 □不知道如何搭建
5	能够创建并运行 Android 程序	□能够创建并运行程序 □明确基本步骤但执行起来存在问题 □不知道如何创建并运行
6	能够对 Android 程序进行调试	□能够通过 Logcat 查找错误，并进行断点调试 □知道在哪里查找错误但无法定位错误 □不知道 Logcat 及断点调试的使用方法
评价得分		
学生自评得分（20%）	学习成果得分（60%）	学习过程得分（20%）	模块综合得分

（1）学生自评得分。

每个学习目标有 3 个选项，选择第 1 个选项得 100 分，选择第 2 个选项得 70 分，选择第 3 个选项得 50 分，学生自评得分为各项学习目标得分的平均分。

（2）学习成果得分。

教师根据学生阶段性测试及模块学习成果的完成情况酌情赋分，满分为 100 分。

（3）学习过程得分。

教师根据学生的其他学习过程表现，如到课情况、参与课程讨论等情况酌情赋分，满分为 100 分。

功能模块 1

用户管理功能的实现

> **说在前面**
>
> **小白**：Android Studio 的安装、配置和基本应用我已经很熟练了，现在我们可以开始模块开发了吧！
>
> **师兄笑笑说**：看把你给急的，今天开始我们就正式开始"学习通关"功能模块的实现了。这第一个模块就是大部分 App 中都包含的用户管理模块。在这个模块中，我们主要完成登录功能和个人信息完善功能。在这个模块我们要循序渐进地完成以下 4 个任务。
>
> 任务 1：登录界面的设计与实现。
>
> 任务 2：注册及详细信息界面的设计与实现。
>
> 任务 3：界面间跳转的实现。
>
> 任务 4：用户信息的存储。

功能需求描述

（1）双击 App 图标，首先进入如图 1-1 所示的登录界面。输入正确的手机号码和密码，单击"登录"按钮，跳转到如图 1-2 所示的"我的信息"界面；如果输入的用户名、密码不正确则给出提示信息。

（2）在"我的信息"界面中，可以修改昵称、性别、生日等信息；修改学习状态时，会跳转到如图 1-3 所示的"学习状态"界面。

（3）"我的信息"界面的用户信息能够进行存储，并在打开"我的信息"界面时显示存储的用户信息。

图 1-1　登录界面

图 1-2 "我的信息"界面

图 1-3 "学习状态"界面

学习目标

1. 知识目标

（1）基本控件 Button、TextView、EditText 等的使用。

（2）常用布局的使用。

（3）Activity 生命周期及 Intent 的使用。

（4）Android 样式及主题。

（5）简单存储的应用。

（6）文件存储的应用。

2. 技能目标

（1）综合运用基本控件、常用布局、样式及主题设计并实现界面。

（2）通过 Intent 实现 Activity 之间的调准及数据传递。

（3）通过 SharedPreferences 实现数据的存取功能。

（4）通过文件存储实现数据存储功能。

任务 1　登录界面的设计与实现

◎ 任务描述

根据前面的参考界面，完成登录界面的设计与实现，并实现如果输入手机号码为"13796612345"，密码为"123456"，单击"登录"按钮，弹出提示信息"手机号码及密码正确，登录成功"，否则，提示"手机号码及密码错误"。

任务学习目标

通过本任务需达到以下目标：
➢ 了解 Android 的界面开发方法。
➢ 能够灵活运用线性布局 LinearLayout 实现界面布局。
➢ 能够运用 TextView、Button、EditText、ImageView 等控件组建界面。
➢ 能够正确使用 shape、selector。
➢ 能够正确使用样式及主题。

技术储备

一、Android 的界面开发方法

Android 的界面开发方法

在 Android 应用中，界面是由布局和控件组成的，布局为整个界面搭了一个有一定排列规则的框架，控件按照规则放在框架的相应位置上。因此，在界面的开发中，我们重点关注布局及控件的使用。在 Android 中，界面的开发方法有两种。

（一）使用 xml 文件编写界面布局

这是 Android 推荐使用的界面开发方法，有效地实现了将界面中的布局代码与 Java 代码隔离，使得代码结构更加清晰，便于代码的维护，同时，能够方便地实现用户界面的自适应转换。

采用 xml 文件编写的界面布局，都放在 res – layout 文件夹中，下面是 HelloWorld 应用程序中 activity_main.xml 的布局代码。

```
1   <?xml version = "1.0" encoding = "utf-8"?>
2   <LinearLayout xmlns:android = "http://schemas.android.com/apk/res/android"
3       xmlns:tools = "http://schemas.android.com/tools"
4       android:layout_width = "match_parent"
5       android:layout_height = "match_parent"
6       tools:context = ".MainActivity" >
7       <TextView
8           android:id = "@+id/textview"
9           android:layout_width = "match_parent"
10          android:layout_height = "wrap_content"
11          android:text = "Hello World!"
12          android:textColor = "@color/colorPrimary" />
13  </LinearLayout>
```

上述代码中，定义了一个线性布局 LinearLayout，在该布局中定义了一个 TextView 控件，在 TextView 控件上显示 HelloWorld。

（二）在 Java 代码中编写布局

除了使用 xml 文件编写布局，也可以在 Java 代码中编写。布局 LinearLayout、控件 TextView 这些 UI（User Interface，用户界面）元素实际都是类名，LinearLayout 继承自 View-

Group，TextView 继承自 View，View 也是所有 UI 元素的父类。ViewGroup 作为容器盛放界面中的控件，它可以包含普通的 View 控件，也可以包含 ViewGroup，它们的包含关系如图 1 – 4 所示。

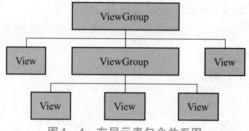

图 1 – 4　布局元素包含关系图

需要注意的是，Android 应用的每个界面根元素必须有且只有一个 ViewGroup 容器。我们可以将上面 xml 文件中的布局使用 Java 代码写在 onCreate( )方法中，如下所示。

```java
public class MainActivity extends AppCompatActivity {
    @Override
    protected void onCreate(Bundle savedInstanceState) {
        super.onCreate(savedInstanceState);
        //创建线性布局对象
        LinearLayout linearLayout = new LinearLayout(this);
        //创建LayoutParams对象,定义了LinearLayout的宽高
        LinearLayout.LayoutParams params = new LinearLayout.LayoutParams(LinearLayout.LayoutParams.MATCH_PARENT,
                LinearLayout.LayoutParams.MATCH_PARENT);
        //创建TextView控件对象,用来显示HelloWorld
        TextView textView = new TextView(this);
        //设置TextView的文字内容
        textView.setText("Hello World!");
        //设置TextView的文字颜色
        textView.setTextColor(getResources().getColor(R.color.colorPrimary));
        //向LinearLayout中添加TextView对象及属性设置
        linearLayout.addView(textView,params);
        //设置在Activity中显示LinearLayout
        setContentView(linearLayout);
    }
}
```

在上述代码中，第 6 ~ 16 行创建了线性布局 LinearLayout 以及 TextView 对象，并进行了相应的设置，第 19 行将 TextView 对象加入布局中，第 21 行将布局设置到界面上。

Android 常用布局及公有属性

二、基本界面布局——线性布局（LinearLayout）

为了适应不同的界面风格，Android 系统提供了多种常用布局，分别为 LinearLayout（线性布

局)、RelativeLayout(相对布局)、FrameLayout(帧布局)、ConstraintLayout(约束布局),这些布局都有各自的控件排列规则,在本任务中我们重点介绍 LinearLayout(线性布局)的使用。

(一)排列规则

线性布局内的子控件呈水平或垂直排列,默认情况下,采用水平排列。需要注意的是,线性布局中控件不会自动换行,排列到屏幕边缘之后,剩下的组件不会被显示出来。

(二)常用的通用属性

在前面的 xml 布局文件中,我们已经见过线性布局的使用方法。其中,xmlns 为 xml 文件的命名空间,android:layout_width 和 android:layout_height 为布局宽和高的属性设置。通过属性可以设置布局的宽、高、背景、边距等。由于布局均为 ViewGroup 的直接或间接子类,ViewGroup 及其父类 View 中定义的属性在布局中均可使用,可以看作布局的通用属性,如图 1-5 所示。

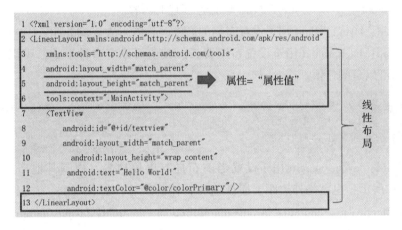

图 1-5 线性布局及其通用属性

常用的布局通用属性见表 1-1。

表 1-1 常用的布局通用属性

属性名称	功能描述
android:id	设置布局的标识名
android:layout_width	设置布局的宽度
android:layout_height	设置布局的高度
android:background	设置布局的背景
android:layout_margin	设置当前布局与屏幕边界或与周围控件的间距
android:padding	设置当前布局与该布局中控件的间距
android:gravity	设置控件中内容相对于控件的对齐方式
android:layout_gravity	设置控件相对于父控件的对齐方式

（1）android：id 用于设置当前布局的唯一标识。在 xml 文件中它的属性值通常通过"@+id/属性名称"定义，为布局指定 android：id 属性后，在自动生成的 R.java 文件中，会生成对应的 int 值。在 Java 代码中通过为 findViewById()方法传入该 int 值来获取该布局对象。

> **注意**：如果属性值中没有 +，则不会在 R 文件中自动生成。这种情况通常用在引用某个控件的时候。

（2）android：layout_width 用于设置布局的宽度，其值可以是具体的尺寸，如 20 dp，也可以是系统定义的值。

①match_parent 表示该布局的宽度与父容器的宽度相同。在前面的代码中，线性布局是界面的根节点，它的父容器宽度就是屏幕的宽度；TextView 文本框的父容器是线性布局，它的父容器宽度就是线性布局的宽度。

②wrap_content 表示该布局的宽度恰好能包裹它的内容。

（3）android：layout_height 用于设置布局的高度，具体用法与 android：layout_width 相同。

（4）android：background 用于设置布局背景，其值可以有多种形式，可以是图片资源、颜色资源，也可以是颜色值。

```
1  android:background = "@drawable/ic_launcher_background"
2  android:background = "@color/colorPrimary"
3  android:background = "#00FF00"
```

（5）android：layout_margin 用于设置当前布局与屏幕边界、周围其他布局或控件的距离，如图 1-6 所示，也就是外边距。属性值为具体的尺寸，如 50 dp。与之相似的还有 android：layout_marginTop、android：layout_marginBottom、android：layout_marginLeft、android：layout_marginRight 属性，分别用于设置当前布局与控件的上、下、左、右边界的距离。

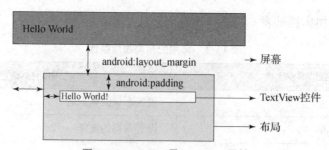

图 1-6 margin 及 padding 属性

（6）android：padding 用于设置当前布局内控件与该布局的距离，也就是内边距，其值可以是具体的尺寸，如 30 dp。与之相似的有 android：paddingTop、android：paddingBottom、android：paddingLeft、android：paddingRight，分别用于设置当前布局中控件与该布局上、下、左、右的距离。

> **注意**：布局必须设置 android：layout_width 和 android：layout_height 属性，其他属性可以根据需求进行设置。

（7）android：gravity 用于设置布局中的内容相对于控件的对齐方式，其值可以为 bottom、top、left、right、center 等。如果设置 LinearLayout 为 android：gravity ="bottom"，表示线性布局中包含的所有控件相对于线性布局底部对齐。

（8）android：layout_gravity 用于设置控件相对于父控件的对齐方式，其取值与 gravity 相同，但表示的含义不同。如果将 LinearLayout 的 android：layout_gravity 属性值设置为 bottom，则表示线性布局在屏幕或者包含它的布局的底部。

（三）特有属性

除以上通用属性外，LinearLayout 还有两个常用的特有属性，见表 1－2 所示。

线性布局 linearlayout 的使用

表 1－2　常用的特有属性

属性名称	功能描述
android：orientation	设置线性布局内控件的排列顺序
android：layout_weight	设置布局内控件的权重

（1）android：orientation 用于设置 LinearLayout 布局中控件的排列顺序，其可选值有 "vertical" 和 "horizontal"。默认情况下，该属性的取值为 "horizontal"。

①vertical 表示布局中的控件从上至下依次竖直排列。

②horizontal 表示布局中的控件从左至右依次水平排列。

（2）android：layout_weight 用于设置将剩余空间按照权重对控件进行分配，该属性取值通常为整数。通过设置该属性值，可以控制布局内的控件按权重比显示大小。比如，两个控件的宽度均为 "wrap_content"，并分别将 "android：layout_weight" 属性值设为 1 和 2，则第 1 个控件将分配剩余空间的 1/3，第 2 个控件将分配剩余空间的 2/3。

【案例 1－1】　编写代码实现如图 1－7 所示的界面效果。

1. 案例分析

（1）界面中包含 3 个水平排列的按钮。

（2）按钮位于界面底部。

（3）控件与屏幕有一定的边距。

2. 实现步骤

（1）创建名为 "LinearLayoutDemo" 的程序。

（2）在 res/layout/activity_main.xml 中编写布局代码。

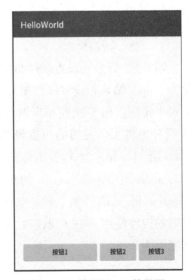

图 1－7　案例 1－1 效果图

整体为水平线性布局，其中，放置 3 个 Button 控件，根据界面分析，需为 LinearLayout 设置 "android：orientation" "android：gravity" "android：layout_margin" 等属性；需为按钮控件设置 "android：layout_weight" 等属性。

```xml
1  <?xml version="1.0" encoding="utf-8"?>
2  <LinearLayout xmlns:android="http://schemas.android.com/apk/res/android"
3      xmlns:tools="http://schemas.android.com/tools"
4      android:layout_width="match_parent"
5      android:layout_height="match_parent"
6      android:layout_margin="10dp"
7      android:padding="5dp"
8      android:gravity="bottom"
9      android:orientation="horizontal">
10     <Button
11         android:layout_width="0dp"
12         android:layout_height="wrap_content"
13         android:layout_weight="2"
14         android:text="按钮1"/>
15     <Button
16         android:layout_width="0dp"
17         android:layout_height="wrap_content"
18         android:layout_weight="1"
19         android:text="按钮2"/>
20     <Button
21         android:layout_width="0dp"
22         android:layout_height="wrap_content"
23         android:layout_weight="1"
24         android:text="按钮3"/>
25 </LinearLayout>
```

第6、7行分别设置线性布局与屏幕的边距为10 dp，线性布局中的按钮控件与布局的间距为5 dp。第8行设置线性布局内的控件位于线性布局底部。第9行设置线性布局内的控件为水平排列。由于线性布局默认排列顺序即为水平排列，也可省略本行代码。第13、18、23行分别将3个按钮的权重设为2、1、1，则3个按钮占据布局的宽度分别为2/4、1/4、1/4。第11、16、21行分别将3个按钮的宽度都设为了0 dp，这是Google官方推荐的写法。因为android：layout_weight的真实含义是设置该属性的控件宽度等于原有宽度（android：layout_width的值）加上剩余空间的占比。为了能够让控件按照权重比例分配大小，当水平排列的控件使用weight属性时，将宽度设为0 dp；当垂直排列的控件使用weight属性时，将高度设为0 dp。

> **注意**：当控件为水平排列时，通常LinearLayout的宽度设置为match_parent，否则android：layout_weight设置有可能失效。同时，水平排列的线性布局中，要注意控件的宽度不要设为match_parent，否则会造成其他控件无法显示。

三、简单控件的使用

Android 提供了非常丰富的界面控件，通过控件可以方便地进行用户界面的开发。接下来将介绍几种简单控件的使用。

TextView 控件的使用

（一）TextView

1. 作用

TextView 控件用于显示文本信息。

2. 继承关系及常用属性

我们可以通过在 xml 文件中为 TextView 添加属性控制 TextView 的显示效果，如图 1 – 8 所示。在 API 文档中有关于 TextView 属性的详细介绍，可以通过查阅 API 文档对 TextView 的继承关系及常用属性进行学习。

图 1 – 8　TextView 显示效果

查文档找答案

查找 API 文档学习 TextView 的继承关系及常用属性。

- TextView 的父类是_____。
- TextView 的子类有_____。
- 根据属性功能描述，查找属性，完成表 1 – 3。

表 1 – 3　Text View 常用属性

属性名称	属性值	功能描述
		设置文本内容
		设置文本颜色
		设置文字大小
		设置文本样式
		设置文本内容的位置
		设置文本最大长度，超出此长度不显示
		设置文本的行数，超出此行数不显示
		设置当文本超出文本框规定范围的显示方式

> **注意**：控件的每一个 xml 属性都对应一个 Java 方法，比如，android：text 属性对应的是 setText( ) 方法，其对应关系也可以在 API 文档中查找。

（二）Button

1. 作用。

Button 控件表示按钮，可以显示文本、图片，同时，可以通过单击来实现交互操作。Button 显示效果如图 1-9 所示。

图 1-9 Button 显示效果

2. 继承关系及常用属性。

Button 继承自 TextView 控件，TextView 中定义的属性在 Button 中均可使用。

3. 单击事件处理。

通常，所有控件都可以设置单击事件，Button 控件就是最为典型及常用的一个。为 Button 控件设置单击事件主要有 3 种方式。

（1）onClick 属性。

在 xml 布局文件中，通过为 Button 控件指定 onClick 属性的值来设置 Button 控件的单击事件。具体单击事件的代码需要在 Activity 中定义一个方法，onClick 属性的值即为方法名。

xml 文件中代码如下。

```
1  <Button
2     android:layout_width = "wrap_content"
3     android:layout_height = "wrap_content"
4     android:onClick = "click"
5     android:text = "按钮1"
6  />
```

Java 文件中代码如下。

```
1  public void click(){
2     //单击按钮需要完成的操作
3  }
```

（2）匿名内部类。

在 Activity 中，使用匿名内部类的方式为 Button 控件设置单击事件。

```
1  //通过 Button 的 ID 获取 Button 对象
2  Button button = this.findViewById(R.id.btn_two);
3  button.setOnClickListener(new View.OnClickListener() {
4     @Override
5     public void onClick(View view) {
6        //单击事件代码
7     }
8  });
```

（3）实现 OnClickListener 接口。

让当前 Activity 实现 OnClickListener 接口，并实现 onClick()方法，Button 对象调用 setOnClickListener()方法实现单击事件监听。

```java
1  public class MainActivity extends AppCompatActivity implements
2                                              View.OnClickListener{
3      @Override
4      protected void onCreate(Bundle savedInstanceState){
5          super.onCreate(savedInstanceState);
6          setContentView(R.layout.activity_main);
7          Button button = this.findViewById(R.id.btn_two);
8          button.setOnClickListener(this);}//设置单击事件监听
9      @Override
10     public void onClick(View view){
11         //实现单击事件的代码
12     }
13 }
```

【案例1-2】 编写代码实现如图1-10所示的界面效果,单击按钮,HelloWorld中的文字会变为相对应的颜色。

1. 案例分析

(1) 界面中包含垂直排列的TextView和4个Button控件。

(2) 控件垂直方向居中。

(3) 4个Button控件均需设置单击事件。

2. 实现步骤

(1) 创建名为"ButtonDemo"的程序。

(2) 在res/layout/activity_main.xml中编写布局代码。

整体为垂直线性布局,其中放置1个TextView和4个Button控件,根据案例分析,需为LinearLayout设置android：orientation、android：gravity等属性。

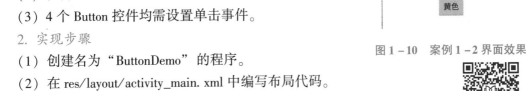

图1-10 案例1-2界面效果

Button的事件处理

```xml
1  <LinearLayout xmlns:android = "http://schemas.android.com/apk/res/android"
2      android:layout_width = "match_parent"
3      android:layout_height = "match_parent"
4      android:orientation = "vertical"
5      android:gravity = "center_horizontal" >
6      <TextView
7          android:id = "@ + id/textView1"
8          android:layout_width = "wrap_content"
9          android:layout_height = "wrap_content"
10         android:text = "Hello World!" />
11     <Button
12         android:id = "@ + id/bt_red"
13         android:layout_width = "wrap_content"
14         android:layout_height = "wrap_content"
15         android:text = "红色" />
16     <Button
17         android:id = "@ + id/bt_blue"
```

```
18        android:layout_width = "wrap_content"
19        android:layout_height = "wrap_content"
20        android:text = "蓝色" />
21     <Button
22        android:id = "@+id/bt_green"
23        android:layout_width = "wrap_content"
24        android:layout_height = "wrap_content"
25        android:text = "绿色" />
26     <Button
27        android:id = "@+id/bt_yellow"
28        android:layout_width = "wrap_content"
29        android:layout_height = "wrap_content"
30        android:text = "黄色" />
31  </LinearLayout>
```

（3）在 MainActivity.java 中编写控制代码。

需获取控件对象，并为 4 个按钮对象添加事件监听代码及单击事件代码。

```
1   public class MainActivity extends AppCompatActivity {
2       private Button bt_red,bt_blue,bt_green,bt_yellow;
3       private TextView textView;
4       @Override
5       protected void onCreate(Bundle savedInstanceState) {
6           super.onCreate(savedInstanceState);
7           setContentView(R.layout.activity_main);
8           init();
9           bt_red.setOnClickListener(new View.OnClickListener() {
10              @Override
11              public void onClick(View v) {
12                  textView.setTextColor(Color.RED);
13              }
14          });
15          bt_blue.setOnClickListener(new View.OnClickListener() {
16              @Override
17              public void onClick(View v) {
18                  textView.setTextColor(Color.rgb(0,0,255));
19              }
20          });
21          bt_green.setOnClickListener(new View.OnClickListener() {
22              @Override
23              public void onClick(View v) {
24                  textView.setTextColor(Color.parseColor("#00FF00"));
25              }
26          });
27          bt_yellow.setOnClickListener(new View.OnClickListener() {
28              @Override
```

```
29            public void onClick(View v) {
30                textView.setTextColor(0xFFFFFF00);
31            }
32        });
33    }
34    public void init() {
35        textView = findViewById(R.id.textView);
36        bt_red = findViewById(R.id.bt_red);
37        bt_blue = findViewById(R.id.bt_blue);
38        bt_green = findViewById(R.id.bt_green);
39        bt_yellow = findViewById(R.id.bt_yellow);
40    }
41 }
```

在 activity_main.xml 中第 4 行设置线性布局为垂直方向，第 5 行设置线性布局中的控件为水平居中对齐。

MainActivity.java 中，第 34 ~ 40 行为所有的控件初始化代码，并在 onCreate()方法的第 4 行调用此方法。

多个按钮事件处理的实现

NainActivity.java 中，第 12、18、24、30 行为 4 种设置文字颜色的方法，分别使用了系统自带颜色类、Color 的静态方法 rgb()、十六进制颜色字符串和 8 位十六进制颜色值。

在上面的代码中，通过匿名内部类的形式分别对 4 个按钮添加了单击事件监听，出现了大量重复代码。对于控件数量较多的界面，在添加单击事件时，更推荐使用实现 OnClickListener 接口的方式，将所有控件的单击事件处理写在 onClick()中，具体代码如下。

```
1  public class MainActivity extends AppCompatActivity implements
2                                           View.OnClickListener {
3      private Button bt_red,bt_blue,bt_green,bt_yellow;
4      private TextView textView;
5      @Override
6      protected void onCreate(Bundle savedInstanceState) {
7          super.onCreate(savedInstanceState);
8          setContentView(R.layout.activity_main);
9          init();
10     }
11     public void init() {
12         textView = findViewById(R.id.textView);
13         bt_red = findViewById(R.id.bt_red);
14         bt_blue = findViewById(R.id.bt_blue);
15         bt_green = findViewById(R.id.bt_green);
16         bt_yellow = findViewById(R.id.bt_yellow);
17         bt_red.setOnClickListener(this);
18         bt_blue.setOnClickListener(this);
19         bt_green.setOnClickListener(this);
20         bt_yellow.setOnClickListener(this);
```

```
21     }
22     @Override
23     public void onClick(View v){
24         switch(v.getId()){
25             case R.id.bt_red:
26                 textView.setTextColor(Color.RED);
27                 break;
28             case R.id.bt_blue:
29                 textView.setTextColor(Color.rgb(0,0,255));
30                 break;
31             case R.id.bt_green:
32                 textView.setTextColor(Color.parseColor("#00FF00"));
33                 break;
34             case R.id.bt_yellow:
35                 textView.setTextColor(0xFFFFFF00);
36         }
37     }
38 }
```

在 onClick()方法中，参数 v 为当前单击的控件对象，获取控件 ID 并与布局中 4 个按钮控件的 ID 进行比较，即可确定当前单击的是哪一个按钮对象，从而进行不同的操作。

> **注意**：对于布局中设置了 ID 属性的控件及资源目录 res 中的资源都会在自动生成的 R 文件中生成一个常量，通过 R. XXXX. XXXX 的形式可以获得。获取控件 ID 的格式为"R. id. 控件 ID"。

（三）EditText

1. 作用

EditText 控件是编辑框，用于输入信息，其显示效果如图 1 - 11 所示。

图 1 - 11 EditText 显示效果

EditText 控件的使用

2. 继承关系及常用属性

我们可以通过在 xml 文件中为 EditText 添加属性，控制 EditText 的显示效果。在 API 文档中有关于 EditText 属性的详细介绍，可以通过查阅 API 文档对 EditText 的继承关系及常用属性进行学习。

查文档找答案

查找 API 文档学习 EditText 的继承关系及常用属性。

● EditText 的父类是_____。

● 根据属性功能描述，查找属性，完成表 1 - 4。

表1-4 Edit Text的常用属性

属性名称	属性值	功能描述
		设置文本内容
		控件中内容为空时显示提示的文本信息
		用于输入密码，隐藏输入内容
		设置输入文本框中的内容只能是数字
		设置文本的最小行数
		设置文本信息超出EditText边界时，是否出现滚动条
		设置是否可编辑
		设置输入类型

以下代码定义了一个提示信息为"这是一个数字编辑框"、输入字体颜色为黑色、字号为20sp，只能输入数字的编辑框。

```
1   <EditText
2       android:layout_width = "match_parent"
3       android:layout_height = "wrap_content"
4       android:hint = "这是一个数字编辑框"
5       android:textColor = "#000000"
6       android:textSize = "20sp"
7       android:inputType = "number" />
```

注意：用于输入密码的输入框，一般通过将android：inputType属性值设置为numberPassword或textPassword实现。

（四）ImageView

1. 作用

ImageView控件可以加载各种图片资源。

ImageView控件的使用

2. 继承关系及常用属性

我们可以通过在xml文件中为ImageView添加属性，控制ImageView的显示效果。在API文档中有关于ImageView属性的详细介绍，可以通过查阅API文档对ImageView的继承关系及常用属性进行学习。

查文档找答案

查找API文档学习ImageView的继承关系及常用属性。

● ImageView的父类是_____。

● ImageView的子类有_____。

● 根据属性功能描述，查找属性，完成表1-5。

表1-5 ImageView 的常用属性

属性名称	属性值	功能描述
		设置 ImageView 控件的背景
		设置 ImageView 控件需要显示的图片资源
		设置图片资源的缩放或移动类型
		设置 ImageView 的最大高度
		设置 ImageView 的最大宽度

以下代码段为两个 ImageView 控件,分别通过 android:background 和 android:src 设置了图片,如图 1-12、图 1-13 所示。从运行结果可以看出,android:background 属性用于设置背景,图片会根据 ImageView 控件的大小进行伸缩;android:src 属性则以原图进行显示,根据实际需要也可以通过 android:scaleType 属性设置图片缩放类型。

```
1  <ImageView
2  android:layout_width="match_parent"
3  android:layout_height="match_parent"
4  android:src=
5  "@drawable/ic_launcher_background"/>
```

```
1  <ImageView
2  android:layout_width="match_parent"
3  android:layout_height="match_parent"
4  android:background=
5  "@drawable/ic_launcher_background"/>
```

【案例1-3】 编写代码实现如图 1-14 所示的简易计算器布局效果。

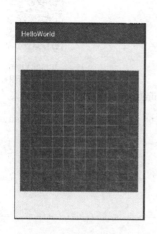 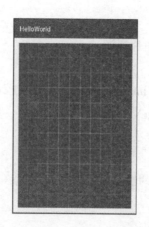 

图 1-12 src 设置　　　图 1-13 background 设置　　　图 1-14 计算器界面效果

1. 案例分析

（1）界面中包含 TextView、EditText、Button 等多种控件。其中，根号为图片，需要使用 ImageView 的子类——ImageButton。

（2）界面分为多行，整体为垂直排列，每一行控件成水平排列，可以通过线性布局嵌套实现整体界面布局。

2. 实现步骤

（1）创建名为"Calculator"的程序。

（2）在 res/layout/activity_main.xml 中编写布局代码。

整体为垂直线性布局，按钮部分每一行为一个水平线性布局，形成线性布局嵌套的布局。通过在每行的 LinearLayout 中设置 android：weight 属性使行与行之间均匀排布，同样，一行中可以通过给 Button 控件设置 android：weight 属性使得按钮均匀排列。

```
1  <?xml version="1.0" encoding="utf-8"?>
2  <LinearLayout xmlns:android="http://schemas.android.com/apk/res/android"
3      xmlns:tools="http://schemas.android.com/tools"
4      android:layout_width="match_parent"
5      android:layout_height="match_parent"
6      android:padding="5dp"
7      android:gravity="top|center"
8      android:orientation="vertical"
9      tools:context=".MainActivity">
10     <TextView
11         android:layout_width="match_parent"
12         android:layout_height="wrap_content"
13         android:gravity="center"
14         android:text="简单计算器"
15         android:textColor="#000000"
16         android:textSize="22sp" />
17     <EditText
18         android:id="@+id/tv_result"
19         android:layout_width="match_parent"
20         android:layout_height="0dp"
21         android:gravity="right|bottom"
22         android:textSize="25sp"
23         android:layout_weight="1" />
24     <LinearLayout
25         android:layout_width="match_parent"
26         android:layout_height="0dp"
27         android:orientation="horizontal"
28         android:layout_weight="1" >
29         <Button
30             android:id="@+id/btn_cancel"
31             android:layout_width="0dp"
```

```xml
32            android:layout_height = "match_parent"
33            android:layout_weight = "1"
34            android:textSize = "30sp"
35            android:text = "CE" />
36        <Button
37            android:id = "@ + id/btn_divide"
38            android:layout_width = "0dp"
39            android:layout_height = "match_parent"
40            android:layout_weight = "1"
41            android:textSize = "30sp"
42            android:text = " ÷ " />
43        <Button
44            android:id = "@ + id/btn_multiply"
45            android:layout_width = "0dp"
46            android:layout_height = "match_parent"
47            android:layout_weight = "1"
48            android:textSize = "30sp"
49            android:text = " × " />
50        <Button
51            android:id = "@ + id/btn_clear"
52            android:layout_width = "0dp"
53            android:layout_height = "match_parent"
54            android:layout_weight = "1"
55            android:textSize = "30sp"
56            android:text = "C" />
57    </LinearLayout>
58    <!-- 省略 第2~3行按钮,代码与第1行一致 -->
59    <LinearLayout
60        android:layout_width = "match_parent"
61        android:layout_height = "0dp"
62        android:orientation = "horizontal"
63        android:layout_weight = "1" >
64        <!-- 省略 第4行前3个按钮,代码与前面一致 -->
65        <ImageButton
66            android:id = "@ + id/ib_sqrt"
67            android:layout_width = "0dp"
68            android:layout_height = "match_parent"
69            android:layout_weight = "1"
70            android:scaleType = "centerInside"
71            android:src = "@drawable/sqrt"
72            android:background = "@drawable/btn_nine_selector"/>
73    </LinearLayout>
74    <LinearLayout
75        android:layout_width = "match_parent"
76        android:layout_height = "0dp"
77        android:orientation = "horizontal"
```

```
 78                android:layout_weight = "1" >
 79            <Button
 80                android:id = "@+id/btn_zero"
 81                android:layout_width = "0dp"
 82                android:layout_height = "match_parent"
 83                android:textSize = "30sp"
 84                android:layout_weight = "2"
 85                android:text = "0" />
 86            <Button
 87                android:id = "@+id/btn_dot"
 88                android:layout_width = "0dp"
 89                android:layout_height = "match_parent"
 90                android:layout_weight = "1"
 91                android:textSize = "30sp"
 92                android:text = "." />
 93            <Button
 94                android:id = "@+id/btn_equal"
 95                android:layout_width = "0dp"
 96                android:layout_height = "match_parent"
 97                android:layout_weight = "1"
 98                android:textSize = "30sp"
 99                android:text = " = " />
100        </LinearLayout>
101    </LinearLayout>
```

长知识:

网格布局（GridLayout）

GridLayout 是 Android 4.0 以后引入的布局，它把整个页面分成 m 行和 n 列，形成 m×n 个单元格的形式。每个单元格都有一个坐标，行坐标为 0~m，列坐标为 0~n。在向网格布局中添加控件时，可以指定控件所在的行坐标和列坐标，也可以不指定，则控件按照从第 1 行第 1 个位置开始，依次按行的顺序进行控件添加。常用属性如下。

- android: columnCount 设置网格列数。
- android: rowCount 设置网格行数。
- android: layout_column 设置控件的列坐标。
- android: layout_row 设置控件的行坐标。
- android: layout_columnSpan 设置控件横向跨的列数。
- android: layout_rowSpan 设置控件纵向跨的行数。

使用网格布局也可以方便地完成计算器界面效果，请你尝试完成。

四、shape、selector 的使用

在界面开发中，我们经常需要实现控件在不同的状态下显示不同的背景或图片的变化效果，比如，在计算器界面中按钮单击前背景为灰色，单击时背景变为绿色，或者编辑框获取焦点前无边框，获取焦点后出现红色边框等动态切换效果。shape 形状和 selector 选择器可以

帮助我们实现这些效果。

（一）shape 的使用

shape 是基础的形状定义工具。用 shape 定义的形状保存为 xml 文件的形式，一般存放在 drawable 目录下。若项目中没有 drawable 目录，则需要新建一个。

使用 shape 可以定义如下 4 种类型的形状。

- rectangle：矩形，默认的形状，可以延伸画出直角矩形、圆角矩形等形状。
- oval：椭圆形，可以延伸画出正圆。
- line：线形，可以定义线形为实线或虚线，以及虚线段的间隔等。
- ring：环形，可以用来画出环形进度条。

在形状定义时，shape 为根节点，可以通过 android：shape 属性指定形状的类型。在 shape 节点下定义了 6 个节点：corners（圆角）、gradien（渐变）、padding（间隔）、size（尺寸）、solid（填充）、stroke（描边）。其中常用的有 corners、solid 及 stroke。

下面，我们创建一个填充为白色、灰色实线描边的圆角矩形，并将其应用在计算器界面中显示数字的 TextView 上，具体步骤如下。

（1）在 res/drawable 目录上右击，依次选择 New→Drawable Resource File，将 Root element 中的 selector 改为"shape"，并填入 File name 为"shape_white_with_stroke"。

（2）在 shape_white_with_stroke 中加入形状设置。

```
1   <? xml version = "1.0" encoding = "utf - 8"? >
2   <shape xmlns:android = "http://schemas.android.com/apk/res/android"
3                        android:shape = "rectangle" >
4       <!--solid 设置形状填充颜色 -->
5       <solid android:color = "#ffffff" />
6       <!--stroke 设置形状描边,width 为描边宽度,此处设置为实线描边
7       如果要设置虚线描边,需要设置 dashGap(虚线的间隔)和 dashWidth(虚线的宽度)
8       两个属性,有一个为 0,则为实线描边 -->
9       <stroke
10          android:width = "1dp"
11          android:color = "#bbbbbb" />
12      <!--corners 设置圆角,圆角半径为 10 dp,只适用于 rectangle -->
13      <corner android:radius = "10dp" />
14  </shape>
```

（3）在 TextView 上设置"shape"作为背景。

```
1   <EditText
2           android:id = "@+id/tv_result"
3           android:layout_width = "match_parent"
4           android:layout_height = "0dp"
5           android:gravity = "right|bottom"
6           android:textSize = "25sp"
7           android:background = "@drawable/shape_white_with_stroke"
8           android:layout_weight = "1"/>
```

（二）selector 的使用

selector 选择器用于为不同的控件状态设置不同的图片背景或颜色变化。它分为"drawable selector"和"color selector"两种，分别为图片选择器和颜色选择器。

1. drawable selector（图片选择器）

图片选择器 xml 文件一般位于 res/drawable 目录下，以下为一个名为"btn_selector.xml"的图片选择器的具体代码，设置了当按钮被按压时及默认状态下的不同显示效果。

```
1  <? xml version = "1.0" encoding = "utf - 8"?>
2  <selector xmlns:android = "http://schemas.android.com/apk/res/android">
3      <!-- 按压时 -->
4      <item android:drawable = "@drawable/button_pressed"
5                            android:state_pressed = "true" />
6      <!-- 默认时 -->
7      <item android:drawable = "@drawable/button_normal" />
8  </selector>
```

其中，每个 item 代表一种控件状态及其对应的图片资源，如 android：state_pressed = "true"代表控件被单击时的状态，android：drawable = "@drawable/button_pressed"代表此时控件的对应图片资源为"button_pressed"。

> **注意：**
> ● @drawable/后的可以是图片资源名，也可以是"shape"，但必须是已经在 res 目录中存在的。
> ● 第 7 行 item 中仅有图片资源，没有状态，表明这是控件默认状态下的情况，默认状态下的设置必须作为 selector 的最后一条。

该 selector 可以通过作为"android：background"等属性的值设置到控件上。这里，我们将其设置到计算器的按钮上，运行后，将出现如图 1 - 15 和图 1 - 16 所示的背景颜色效果。

```
1  <Button
2          android:id = "@ + id/btn_cancel"
3          android:layout_width = "0dp"
4          android:layout_height = "match_parent"
5          android:layout_weight = "1"
6          android:textSize = "30sp"
7          android:background = "@drawable/selector_button"
8          android:text = "CE" />
```

图 1 - 15　按钮未被单击时效果

图 1 - 16　按钮被单击时效果

2. color selector（颜色选择器）

颜色选择器 xml 文件一般位于 res/color 目录下，当我们想要实现控件上的文字在不同状态下显示不同颜色的时候，通常会用到颜色选择器。以下为一个名为"text_color_selector.xml"的图片选择器的具体代码。

```
1  <?xml version = "1.0" encoding = "utf-8"?>
2  <selector xmlns:android = "http://schemas.android.com/apk/res/android">
3      <!-- 按压时 -->
4      <item android:color = "#ffffff" android:state_pressed = "true" />
5      <!-- 默认时 -->
6      <item android:color = "#000000" />
7  </selector>
```

上述代码中第 4、6 行分别对应控件按下及默认情况下的颜色值。该 selector 可以通过"android：textColor"等可以设置颜色的属性值设置到控件上。运行后，将出现如图 1-15 和图 1-16 所示的文字颜色效果。

```
1  <Button
2          android:id = "@+id/btn_cancel"
3          android:layout_width = "0dp"
4          android:layout_height = "match_parent"
5          android:layout_weight = "1"
6          android:textSize = "30sp"
7          android:background = "@drawable/selector_button"
8          android:textColor = "@color/text_color_selector"
9          android:text = "CE" />
```

> **长知识：**
>
> **控件状态还有哪些？**
>
> 除了上面提到的控件按压状态外，控件还有以下状态，通过在 selector 选择器中为不同的状态设置不同的图片或颜色效果，可以实现更为丰富的界面效果。
>
> - android：state_focused 表示控件是否获取焦点。
> - android：state_selected 表示列表控件中列表项是否被选中。
> - android：state_checkable 表示控件是否有选中状态，一般用于可选择控件。
> - android：state_checked 表示控件是否被选中，一般用于单选、多选控件。
> - android：state_hovered 表示光标是否移动到控件之上。
> - android：state_window_focused 表示当前界面是否获得焦点。

五、样式和主题

在前面计算器界面的布局中，我们会发现一个问题，每个按钮的属性设置基本相同，由于按钮众多，这些属性设置我们需要重复写很多遍，并且当属性设置需要发生变化的时候，需要多处重复修改，这给程序的修改和维护带来很多困难。这个问题可以通过样式进行优化。

（一）style（样式）

样式是一系列对于控件具体属性的设置。通过创建样式，可以方便地对 View 控件进行美化。Android 的样式一般定义在 "res/values/styles.xml" 文件中，其中有一个根元素 <resources>，而具体的每种样式定义则是通过 <resources> 下的子标签 <style> 来完成的，style 标签中通过 name 属性定义样式名，在 <style> 标签中可以添加多个 <item> 来设置样式不同的属性。具体示例代码如下。

```
1  <resources>
2  <style name="btnStyle">
3      <item name="android:layout_width">50dp</item>
4      <item name="android:layout_height">50dp</item>
5      <item name="android:textColor">#000000</item>
6  </style>
7  </resources>
```

在布局文件中的 View 控件可通过 style 属性调用样式，其中，属性值 "@style/" 之后为样式名，示例代码如下。

```
1  <Button
2      ……
3      style="@style/btnStyle"
4  </Button>
```

（二）theme（主题）

样式能够对控件进行美化，而主题则针对窗体的样式，能够对整个应用或者某个 Activity 产生全局应用。主题资源同样在 "res/values/styles.xml" 文件中定义，示例代码如下。

```
1  <resources>
2  <!-- Base application theme. -->
3  <style name="AppTheme" parent="Theme.AppCompat.Light.DarkActionBar">
4      <!-- Customize your theme here. -->
5      <item name="colorPrimary">@color/colorPrimary</item>
6      <item name="colorPrimaryDark">@color/colorPrimaryDark</item>
7      <item name="colorAccent">@color/colorAccent</item>
8  </style>
9  </resources>
```

主题的定义与 style 的定义格式基本一致，此时 name 值为主题名，parent 属性用于指定当前主题的父主题为 Android 系统提供的 "Theme.AppCompat.Light.DarkActionBar"，item 为当前主题的具体样式。

要设置主题，一般在 AndroidManifest.xml 中设置，示例代码如下。

```
1  <application
2      ……
3      android:theme="@style/AppTheme">
4  </application>
```

【案例1-4】 使用样式对计算器界面布局代码进行优化。

1. 案例分析

在计算器界面代码中，每个按钮的属性设置基本相同，可以将属性及属性值写入样式，在控件中调用样式。

2. 实现步骤

（1）在 res/values 目录下的 styles 文件中，加入名为"btnStyle"的样式。

```
1   <resources>
2       <style name = "btnStyle">
3           <item name = "android:layout_width">0dp</item>
4           <item name = "android:layout_height">match_parent</item>
5           <item name = "android:layout_weight">1</item>
6           <item name = "android:gravity">center</item>
7           <item name = "android:textColor">@color/text_color_
8   selector</item>
9           <item name = "android:textSize">30sp</item>
10          <item name = "android:background">@drawable/selector_
11  button</item>
12      </style>
13  </resources>
```

（2）将 activity_main.xml 中所有按钮公共属性去掉，替换为 style = "@style/btnStyle"。

```
1   <?xml version = "1.0" encoding = "utf-8"?>
2   <LinearLayout xmlns:android = "http://schemas.android.com/apk/res/android"
3       xmlns:tools = "http://schemas.android.com/tools"
4       android:layout_width = "match_parent"
5       android:layout_height = "match_parent"
6       android:padding = "5dp"
7       android:gravity = "top|center"
8       android:orientation = "vertical"
9       tools:context = ".MainActivity">
10      <TextView
11          android:layout_width = "match_parent"
12          android:layout_height = "wrap_content"
13          android:gravity = "center"
14          android:text = "简单计算器"
15          android:textColor = "#000000"
16          android:textSize = "22sp" />
17      <TextView
18          android:id = "@+id/tv_result"
19          android:background = "@drawable/shape_white_with_stroke"
20          android:layout_width = "match_parent"
21          android:layout_height = "0dp"
22          android:gravity = "right|bottom"
23          android:lines = "3"
24          android:maxLines = "3"
```

```xml
25        android:scrollbars = "vertical"
26        android:textColor = "#000000"
27        android:textSize = "25sp"
28        android:layout_weight = "1" />
29    <LinearLayout
30        android:layout_width = "match_parent"
31        android:layout_height = "0dp"
32        android:orientation = "horizontal"
33        android:layout_weight = "1" >
34        <Button
35            android:id = "@+id/btn_cancel"
36            style = "@style/btnStyle"
37            android:text = "CE" />
38        <Button
39            android:id = "@+id/btn_divide"
40            style = "@style/btnStyle"
41            android:text = " ÷ " />
42        <Button
43            android:id = "@+id/btn_multiply"
44            style = "@style/btnStyle"
45            android:text = " × " />
46        <Button
47            android:id = "@+id/btn_clear"
48            style = "@style/btnStyle"
49            android:text = "C" />
50    </LinearLayout>
51    <!--省略第2、3、4行代码 -->
52    <LinearLayout
53        android:layout_width = "match_parent"
54        android:layout_height = "0dp"
55        android:orientation = "horizontal"
56        android:layout_weight = "1" >
57        <Button
58            android:id = "@+id/btn_zero"
59            style = "@style/btnStyle"
60            android:layout_weight = "2"
61            android:text = "0" />
62        <Button
63            android:id = "@+id/btn_dot"
64            style = "@style/btnStyle"
65            android:text = "." />
66        <Button
67            android:id = "@+id/btn_equal"
68            style = "@style/btnStyle"
69            android:text = " = " />
70    </LinearLayout>
71 </LinearLayout>
```

注意，第 57~61 行的 Button 控件中，btnStyle 样式里已经包含了对 android：layout_weight 属性的设置，其值设置为 1，而第 60 行又加入了 android：layout_weight = "2" 的属性设置。此时，控件中设置的属性值会替代样式中的属性值。

> **长知识：**
>
> **Toast 的使用**
>
> Toast 主要用于显示提示信息，其显示方式为在屏幕下方浮现出一个窗口，显示一段时间后又消失。Toast 的基本用法如下。
>
> Toast. makeText(MainActivity. this,'要显示的内容', Toast. LENGTH_SHORT). show( );
>
> 其中，makeText( ) 为 Toast 的静态方法，第 1 个参数为上下文对象，可用 getApplicationContext( ) 或 getContext( ) 或 this；第 2 个参数为提示信息内容；第 3 个参数为显示时间长短，其值为常量。
>
> 最后，不要忘记调用 show( ) 方法，不然 Toast 无法显示。

任务实施

1. 任务分析

（1）界面整体布局为垂直方向线性布局，上方横排的两个图标及下方找回密码和账号挂失可采用水平方向线性布局，如图 1 – 17 所示。

（2）按钮被单击、未被单击时发生的背景变化如图 1 – 18 所示，需定义形状及图片选择器。

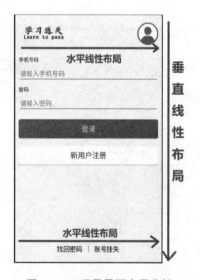

图 1 – 17　登录界面布局分析　　　　　图 1 – 18　按钮背景变化效果

2. 实现步骤

（1）创建名为 "LearnToPass" 的项目，指定包名为 "com. project. learntopass"。

（2）将界面中所需用到的图标 logo. png 和 user_photo. png 放在 res/mipmap – xxhdpi 目录下。

(3) 在 res/drawable 中创建 shape_login_btn_select.xml、shape_login_btn_not_select.xml、shape_register_btn_select.xml、shape_register_btn_not_select.xml 文件,编写"登录"及"注册"按钮背景变化所需的形状。"登录"按钮未被单击时所需形状的具体代码如下,其他形状代码只需修改填充颜色即可。

【代码 1-1】 shape_login_btn_select.xml

```
1  <?xml version = "1.0" encoding = "utf-8"?>
2  <shape xmlns:android = "http://schemas.android.com/apk/res/
3                        android" android:shape = "rectangle">
4      <solid android:color = "@color/red"/> <!-- 填充颜色 -->
5      <corners android:radius = "5dp"/> <!-- 设置圆角 -->
6      <size android:width = "200dp" android:height = "50dp"/>
7      <stroke android:width = "1dp" android:color = "@color/red"
8          android:dashGap = "0dp" android:dashWidth = "0dp"/>
9  </shape>
```

(4) 在 res/drawable 中创建 selector_login_btn.xml、selector_register_btn.xml 文件,编写"登录"及"注册"按钮背景 selector 代码。"登录"按钮 selector 选择器代码如下所示,"注册"按钮相应代码中只需修改 shape 文件名即可。

【代码 1-2】 selector_login_btn.xml

```
1  <?xml version = "1.0" encoding = "utf-8"?>
2  <selector xmlns:android = "http://schemas.android.com/apk/res/android">
3      <!-- 设置不同控件状态下,显示不同的 shape -->
4      <!-- 按压时 -->
5      <item android:state_pressed = "true"
6          android:drawable = "@drawable/shape_login_btn_not_select"/>
7      <!-- 默认时,必须写在所有特殊状态的最后 -->
8      <item android:drawable = "@drawable/shape_login_btn_select"/>
9  </selector>
```

(5) 在 activity_login.xml 中编写登录界面布局代码。

【代码 1-3】 activity_login.xml

```
1   <?xml version = "1.0" encoding = "utf-8"?>
2   <LinearLayout xmlns:android = "http://schemas.android.com/apk/res/android"
3       android:layout_width = "match_parent"
4       android:layout_height = "match_parent"
5       android:orientation = "vertical"
6       android:padding = "10dp">
7       <!-- 顶部实现 开始 -->
8       <LinearLayout
9           android:layout_width = "match_parent"
10          android:layout_height = "80dp"
11          android:orientation = "horizontal">
```

```xml
12            <!-- 标志存放 -->
13            <ImageView
14                android:layout_width = "140dp"
15                android:layout_height = "match_parent"
16                android:layout_gravity = "center"
17                android:src = "@mipmap/logo" />
18            <!-- 中间空白 -->
19            <TextView
20                android:layout_width = "0dp"
21                android:layout_height = "match_parent"
22                android:layout_weight = "1" />
23            <!-- 用户头像 -->
24            <ImageView
25                android:layout_width = "60dp"
26                android:layout_height = "60dp"
27                android:layout_gravity = "center"
28                android:src = "@mipmap/user_photo" />
29        </LinearLayout>
30        <!-- 中部实现 -->
31        <LinearLayout
32            android:layout_width = "match_parent"
33            android:layout_height = "0dp"
34            android:layout_weight = "1"
35            android:orientation = "vertical"
36            android:paddingTop = "20dp" >
37            <!-- 手机号 -->
38            <TextView
39                android:layout_width = "wrap_content"
40                android:layout_height = "wrap_content"
41                android:text = "@string/phone_number"
42                android:textColor = "@color/black"
43                android:textSize = "14sp" />
44            <EditText
45                android:id = "@+id/et_phonenum"
46                android:layout_width = "match_parent"
47                android:layout_height = "wrap_content"
48                android:hint = "请输入手机号码"
49                android:inputType = "number"
50                android:textSize = "18sp"
51                android:layout_marginTop = "5dp" />
52            <!-- 密码 -->
53            <TextView
54                android:layout_width = "wrap_content"
55                android:layout_height = "wrap_content"
56                android:text = "@string/password"
```

```xml
57          android:textColor = "@color/black"
58          android:textSize = "14sp"
59          android:layout_marginTop = "10dp" />
60      <EditText
61          android:id = "@+id/et_password"
62          android:layout_width = "match_parent"
63          android:layout_height = "wrap_content"
64          android:hint = "请输入密码"
65          android:inputType = "textPassword"
66          android:textSize = "18sp"
67          android:layout_marginTop = "5dp" />
68      <!-- 登录按钮和新用户注册按钮 -->
69      <Button
70          android:id = "@+id/btn_login"
71          android:layout_width = "match_parent"
72          android:layout_height = "wrap_content"
73          android:layout_marginTop = "20dp"
74          android:background = "@drawable/selector_login_btn"
75          android:text = "登录"
76          android:textColor = "@color/white"
77          android:textSize = "20sp" />
78      <Button
79          android:id = "@+id/btn_register"
80          android:layout_width = "match_parent"
81          android:layout_height = "wrap_content"
82          android:layout_marginTop = "20dp"
83          android:background = "@drawable/selector_register_btn"
84          android:text = "新用户注册"
85          android:textColor = "@color/red"
86          android:textSize = "20sp" />
87  </LinearLayout>
88  <!-- 底部实现 -->
89  <LinearLayout
90      android:layout_width = "match_parent"
91      android:layout_height = "50dp"
92      android:gravity = "center" >
93      <TextView
94          android:layout_width = "wrap_content"
95          android:layout_height = "wrap_content"
96          android:text = "找回密码"
97          android:textColor = "@color/black"
98          android:textSize = "18sp" />
99      <!-- 竖线 -->
100     <View
```

```
101                 android:layout_width = "2dp"
102                 android:layout_height = "match_parent"
103                 android:layout_margin = "16dp"
104                 android:background = "@color/gray" />
105             <TextView
106                 android:layout_width = "wrap_content"
107                 android:layout_height = "wrap_content"
108                 android:text = "账号挂失"
109                 android:textColor = "@color/black"
110                 android:textSize = "18sp" />
111         </LinearLayout>
112 </LinearLayout>
```

(6) 在 MainActivity.java 中编写界面功能逻辑代码。"登录"按钮单击事件处理代码如【代码 1-4】所示。

【代码 1-4】 MainActivity.java

```
1  public class MainActivity extends AppCompatActivity{
2      private Button btn_login;
3      private EditText et_phonenum,et_password;
4      private String phonenum,password;
5      @Override
6      protected void onCreate(Bundle savedInstanceState) {
7          super.onCreate(savedInstanceState);
8          setContentView(R.layout.activity_main);
9          init();
10         btn_login.setOnClickListener(new View.OnClickListener() {
11             @Override
12             public void onClick(View arg0) {
13                 phonenum = et_phonenum.getText().toString();
14                 password = et_password.getText().toString();
15                 if(phonenum = = null || "".equals(phonenum)){
16                     Toast.makeText(MainActivity.this,"请输入手机号码",
17                         Toast.LENGTH_LONG).show();
18                 }else if(password = = null || "".equals(password)){
19                     Toast.makeText(MainActivity.this,"请输入密码",
20                         Toast.LENGTH_LONG).show();
21                 }else if(phonenum.equals("12345678910") &&
22                             password.equals("123456")){
23                     Toast.makeText(MainActivity.this,"登录成功",
24                         Toast.LENGTH_LONG).show();
25                 }else{
26                     Toast.makeText(MainActivity.this,"手机号码或密码错误",
27                         Toast.LENGTH_LONG).show();
28                 }
29             }
```

```
30              });
31      }
32      public void init() {
33          btn_login = findViewById(R.id.btn_login);
34          et_phonenum = findViewById(R.id.et_phonenum);
35          et_password = findViewById(R.id.et_password);
36      }
37  }
```

> **长知识：**
>
> **如何去掉 Activity 的标题栏？**
>
> 在 Activity 上方会有默认的深色标题栏，出现的原因就是当前主题 AppTheme 的父主题 Theme.AppCompat.Light.DarkActionBar 默认深色标题栏。如果要想将所有应用中 Activity 的标题栏都去掉，则可以将父主题的最后部分改为 NoActionBar。
>
> 如果仅某几个 Activity 不需要标题栏，则可以通过重新定义一个新的 Style 来实现。代码如下所示。
>
> ```xml
> <!-- 去除 ActionBar -->
> <style name="AppTheme.NoActionBar">
>     <item name="windowActionBar">false</item>
>     <item name="windowNoTitle">true</item>
> </style>
> ```
> 在 AndroidManifest.xml 中，相应 Activity 的主题属性设置为 AppTheme.NoActionBar 即可。
>
> ```xml
> <activity android:theme="@style/AppTheme.NoActionBar"/>
> ```

任务反思

编写并运行登录界面，将在代码编写及程序调试过程中出现的异常信息、产生原因及解决方法记录在下方。

问题1：_____

产生原因：_____

解决方法：_____

问题2：_____

产生原因：_____

解决方法：_____

任务总结及巩固

师兄：小白，这部分我们完成了登录界面，你来总结一下这一部分你都学习到了哪些内容？

小白：一个界面由布局和控件组成，我们需要合理使用布局、排列控件位置并设置恰当的属性值来达到界面要求。具体来说，我学会了以下几点。

- 线性布局一线排，要么水平要么垂直。
- 虽说简单易排列，要想复杂可嵌套。
- 文本按钮编辑框，简单控件最常用。
- 形状样式选择器，控件美观作用大。
- 就是属性真不少，想要记清需多练。

师兄：总结得不错！这部分我们主要学习的是界面布局，属性多又易混淆，是需要多下下功夫。

一、基础巩固

1. 如果需要将 Button 按钮上的文字设为"提交"，需使用的属性是（   ）。

   A. android：id   B. android：text
   C. android：background   D. android：gravity

2. 如果需要使线性布局中的 4 个按钮横向排列，下面属性设置正确的是（   ）。

   A. android：orientation = "horizontal"
   B. android：gravity = "center_horizontal"
   C. android：orientation = "vertical"
   D. android：layout_gravity = "center"

3. 在使用 EditText 控件时，如果需要实现当文本内容为空时，显示提示信息，可以使用的属性是（   ）。

   A. android：inputType   B. android：minLines
   C. android：hint   D. android：textSize

4. 在使用 shape 定义基本形状的时候，要设置圆角矩形的圆角半径需添加（   ）节点。

   A. solid   B. stroke   C. corners   D. size

5. 在 ImageView 上设置一张图片，需使用的属性是（   ）。

   A. android：src   B. android：background
   C. android：scaleType   D. android：img

二、技术实践

利用线性布局完成如图 1-19 所示的注册界面。

图 1-19 注册界面

任务 2　注册及详细信息界面的设计与实现

任务描述

参考如图 1-20 所示的界面，完成"我的信息"界面的设计与实现，单击昵称可弹出"修改昵称"对话框进行昵称修改，单击性别可弹出"性别选择"对话框进行选择，单击生日可弹出日期对话框，单击学院后的列表控件可以选择所在学院。

图 1-20 "我的信息"页面及弹出对话框样式

— 67 —

任务学习目标

通过本任务需达到以下目标：
- 能够正确使用 AlertDialog 创建对话框。
- 能够灵活运用相对布局 RelativeLayout 实现界面布局。
- 能够运用 Spinner 等控件组建界面。

技术储备

相对布局 RelativeLayout

一、相对布局——RelativeLayout

RelativeLayout 是 Android 基本布局之一，也是在界面布局中经常会用到的一种布局形式。

（一）排列规则

相对布局中控件的位置以其兄弟控件或父容器作为参照物决定。在进行布局的时候注意：如果 A 的位置由 B 的位置决定，则需要先定义 B 控件，再定义 A 控件。

（二）常用属性

RelativeLayout 中的子控件在定义相对位置时可以使用 15 个属性，根据其参照物不同，可分为 4 组。

查文档找答案

查找 API 文档学习 RelativeLayout 的继承关系及常用属性。

- RelativeLayout 的父类是 _____。
- 根据属性功能描述，查找属性，完成表 1-6。

表 1-6 RelativeLayout 的常用属性

属性名称	功能描述
	设置当前控件位于某控件上方
	设置当前控件位于某控件下方
	设置当前控件位于某控件左侧
	设置当前控件位于某控件右侧
	设置当前控件的上边界与某控件上边界对齐
	设置当前控件的下边界与某控件下边界对齐
	设置当前控件的左边界与某控件左边界对齐
	设置当前控件的右边界与某控件右边界对齐
	设置当前控件是否位于父控件的中央
	设置当前控件是否位于父控件的垂直居中位置

续表

属性名称	功能描述
	设置当前控件是否位于父控件的水平居中位置
	设置当前控件是否与父控件顶端对齐
	设置当前控件是否与父控件左对齐
	设置当前控件是否与父控件右对齐
	设置当前控件是否与父控件底端对齐

> **注意：**
> • 表1–6中前8个属性为相对于其他控件的位置关系，因此，属性值为参照控件的 id，以@id/XXXXX 表示。
> • 后7个属性为相对于父控件的位置关系，属性值为 true/false。

【案例1–5】 编写代码实现如图1–21所示的界面效果。

1. 案例分析

（1）上方的按钮处于界面水平居中的位置，下方按钮处于界面底部。

（2）中间的3个按钮，可以以"中间"按钮为基准，"左方"和"右方"两个按钮分别相对于"中间"按钮进行相对布局。

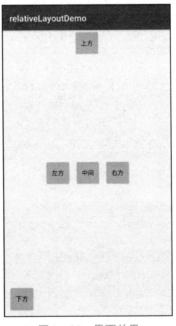

图1–21 界面效果

2. 实现步骤

(1) 创建程序。创建名为"RelativeLayoutDemo"的程序。

(2) 在 res/layout/activity_main.xml 中编写布局代码。

```xml
1   <?xml version="1.0" encoding="utf-8"?>
2   <RelativeLayout xmlns:android="http://schemas.android.com/apk/res/android"
3       xmlns:tools="http://schemas.android.com/tools"
4       android:layout_width="match_parent"
5       android:layout_height="match_parent"
6       tools:context=".MainActivity">
7       <Button
8           android:id="@+id/center"
9           android:layout_width="60dp"
10          android:layout_height="60dp"
11          android:text="中间"
12          android:layout_centerInParent="true"/>
13      <Button
14          android:id="@+id/center_left"
15          android:layout_width="60dp"
16          android:layout_height="60dp"
17          android:text="左方"
18          android:layout_marginRight="10dp"
19          android:layout_toLeftOf="@id/center"
20          android:layout_alignTop="@id/center"/>
21      <Button
22          android:id="@+id/center_right"
23          android:layout_width="60dp"
24          android:layout_height="60dp"
25          android:text="右方"
26          android:layout_marginLeft="10dp"
27          android:layout_toRightOf="@id/center"
28          android:layout_alignTop="@id/center"/>
29      <Button
30          android:id="@+id/center_top"
31          android:layout_width="60dp"
32          android:layout_height="60dp"
33          android:text="上方"
34          android:layout_marginBottom="10dp"
35          android:layout_centerHorizontal="true"/>
36      <Button
37          android:id="@+id/center_bottom"
38          android:layout_width="60dp"
39          android:layout_height="60dp"
40          android:text="下方"
41          android:layout_margin="10dp"
42          android:layout_alignParentBottom="true"/>
43  </RelativeLayout>
```

第 8~13 行定义了中间的按钮，它位于整个界面的中央，使用 android：layout_centerInParent 属性确定了它的基准位置。

第 14~21 行定义了左方的按钮，它以中间按钮为基准，位于中间按钮的左侧，使用 android：layout_toLeftOf = "@id/center" 来确定位置；并且它的上边缘与中间按钮对齐，使用 android：layout_alignTop = "@id/center" 来确定位置。通常，我们可以通过多个相对关系来确定控件的准确位置。

第 30~36 行及第 37~43 行的代码定义了上方和下方的按钮，它们分别通过 android：layout_centerHorizontal 和 android：layout_alignParentBottom 属性设置相对于父控件的位置。

二、列表框控件——Spinner

Spinner 控件的基本使用

（一）作用

Spinner 为列表框控件，用于从一组选项中选择某一项。它有两种展示方式，一种是在当前下拉框正下方显示下拉列表，另一种是以弹出对话框的形式展示列表项。

（二）继承关系及常用属性

Spinner 是 AdapterView（适配器视图）的间接子类。除了控件的常用属性外，我们通常使用 spinnerMode 属性设置 Spinner 的展示方式，当其值为 dropdown 时，在当前下拉框正下方显示下拉列表，其值为 dialog 时，以弹出对话框的形式展示列表项，如图 1-22 所示。

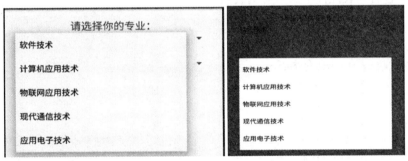

图 1-22　下拉列表及弹出对话框形式的列表框显示效果

（三）使用方法

Spinner 是适配器视图的一个基本控件，在使用时，通常要先在布局中加入 Spinner 控件，再设置列表项数据。Spinner 设置列表项数据有两种方式，一种是通过资源文件进行设置，另一种是通过适配器进行设置。

在页面布局中通常按如下代码加入 Spinner 控件。

```
1    <?xml version = "1.0" encoding = "utf-8"?>
2    <LinearLayout xmlns:android = "http://schemas.android.com/apk/res/android"
3        xmlns:tools = "http://schemas.android.com/tools"
4        android:layout_width = "match_parent"
```

```
5        android:layout_height = "match_parent"
6        android:padding = "20dp"
7        android:orientation = "vertical"
8        tools:context = ".MainActivity" >
9        <TextView
10           android:layout_width = "match_parent"
11           android:layout_height = "wrap_content"
12           android:gravity = "center"
13           android:text = "请选择你的专业:"
14           android:textSize = "20sp" />
15       <Spinner
16           android:id = "@ + id/spinner "
17           android:layout_width = "match_parent"
18           android:layout_height = "wrap_content"
19           android:layout_marginTop = "5dp"
20           android:entries = "@array/major"
21           android:spinnerMode = "dropdown" />
22   </LinearLayout>
```

1. 通过资源文件设置列表项数据

在 res/values 目录下，创建 arrs.xml 资源文件。在文件中定义字符串数组资源，name 为资源名，其中包含列表控件中的所有列表项。

```
1    <? xml version = "1.0" encoding = "utf - 8"? >
2    <resources>
3    <string - array name = "major" >
4         <item>软件技术</item>
5         <item>计算机应用技术</item>
6         <item>物联网应用技术</item>
7         <item>现代通信技术</item>
8         <item>应用电子技术</item>
9    </string - array >
10   </resources>
```

在布局文件中，使用 android：entries 属性，为 Spinner 控件设置列表项数据。其属性值为上面定义的数组资源，写法为"@array/资源名"。

```
1    <Spinner
2        android:id = "@ + id/spinner"
3        android:layout_width = "match_parent"
4        android:layout_height = "wrap_content"
5        android:entries = "@array/major"
6        android:spinnerMode = "dropdown" />
```

使用这种方式设置的列表项通常不经常发生变化，一旦列表项数据发生变化，就需要修改资源文件。如果列表项需要从数据库或其他数据源读取，通常会采用适配器进行数据设置。

2. 通过适配器设置列表项数据

通过适配器设置列表项数据需要在 Java 代码中完成。在 MainActivity.java 中我们通常会按如下几个步骤进行。

①定义列表项数据，可以是数组或集合的形式。
②创建适配器对象，并对适配器进行设置。
③获取 Spinner 对象，并调用如下方法进行 Spinner 对象的设置。

- setPrompt( )：设置标题文字。
- setAdapter( )：设置适配器。
- setSelection( )：设置当前选中项，需在 setAdapter( )方法后调用。

具体代码如下。

使用 Adapter 加载列表项

```
1  public class MainActivity extends AppCompatActivity {
2      private Spinner spinner;
3      private String[] majors;
4      @Override
5      protected void onCreate(Bundle savedInstanceState) {
6          super.onCreate(savedInstanceState);
7          setContentView(R.layout.activity_main);
8          //1. 获取列表框控件
9          spinner = findViewById(R.id.spinner);
10         //2. 准备数据
11         majors = new String[]{"软件技术","计算机应用技术",
12                 "物联网应用技术","现代通信技术","应用电子技术"};
13         //3. 创建 Adapter
14         ArrayAdapter adapter = new ArrayAdapter(MainActivity.this,
15                 android.R.layout.simple_list_item_1,majors)
16         //4. 设置 Adapter
17         spinner.setAdapter(adapter);
18         spinner.setPrompt("请选择你的专业:");
19     }
20  }
```

（四）ArrayAdapter 适配器

在使用 Spinner 时，实际上分为两个部分，一部分是 Spinner，它好比一个巧克力外盒；另一部分是数据，它们好比一块块巧克力。巧克力需要包在包装纸里放到巧克力外盒中展示出来。适配器就好比一个巧克力包装流水线，我们把一块块巧克力（数据）给适配器，它就把巧克力排列好，包裹进提前做好的包装纸（数据列表项布局）中，出来的就是一个个已经包裹着包装纸的巧克力，如图 1-23 所示。

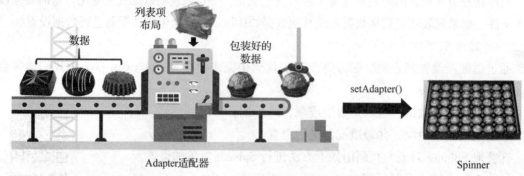

图 1-23 理解适配器

巧克力包装可能是简单的,也可能是比较复杂的,这就对应于不同的流水线(适配器)。

ArrayAdapter 就是最简单的流水线,用于每个列表项只展示文本的情况,上面代码段中第 14~15 行就是它的构造方法的具体用法。

其中,第 1 个参数为上下文对象;第 2 个参数为列表未单击下拉按钮时当前文本的显示布局,这里的 android.R.layout.simple_list_item_1 是 android 的自带布局,其中只有一个 TextView;第 3 个参数为数据数组。

我们也可以调用 setDropDownViewResource() 方法设置单击下拉按钮时下拉列表中列表项的显示布局,具体代码如【案例 1-6】所示,其中 R.layout.item_dropdown 为我们自己写的显示布局。

(五) Spinner 的事件处理

通过调用 setOnItemSelectedListener() 可以为 Spinner 设置下拉列表项选中事件监听器,同时,需要实现 OnItemSelectedListener 接口。

```
1   spinner.setOnItemSelectedListener(new AdapterView.OnItemSelectedListener() {
2       @Override
3       public void onItemSelected(AdapterView<?> adapterView,
4                                   View view,int i,long l) {
5           //事件处理代码
6           Toast.makeText(MainActivity.this,"您的专业为:" + majors[i],
7                           Toast.LENGTH_LONG).show();
8       }
9       @Override
10      public void onNothingSelected(AdapterView<?> adapterView) {}
11  });
```

这里重点关注 onItemSelected() 方法,当有下拉列表项被选中时调用此方法,其第 1 个参数为当前 Spinner 对象,第 2 个参数为单击的列表项,第 3 个参数为选中列表项所在的位置,第 4 个参数选中列表项所在行,与第 3 个参数一致。

【案例 1-6】 编写代码实现如图 1-24 所示的下拉列表效果,选中某项后弹出提示信息。

Spinner 的事件处理

1. 案例分析

本案例实现效果中，下拉列表中列表项的选中效果与选中列表项的实现效果不同，且不同于 Android 自带的列表项效果，因此，需要自定义两种列表项的布局。

2. 实现步骤

（1）编写下拉列表项的布局 item_dropdown.xml 及选中列表项的布局 item_selector.xml。

下面以下拉列表项布局文件 item_dropdown.xml 为例，具体代码如下。

图 1-24 界面效果

```
1  <TextView xmlns:android = "http://schemas.android.com/apk/res/android"
2      android:id = "@ + id/tv_name"
3      android:layout_width = "match_parent"
4      android:layout_height = "40dp"
5      android:singleLine = "true"
6      android:gravity = "center"
7      android:textSize = "17sp"
8      android:textColor = "#ff0000" />
```

（2）编写主界面布局效果。

```
1   <?xml version = "1.0" encoding = "utf-8"?>
2   <LinearLayout xmlns:android = "http://schemas.android.com/apk/res/android"
3       android:layout_width = "match_parent"
4       android:layout_height = "match_parent"
5       android:orientation = "vertical"
6       android:padding = "20dp" >
7       <Spinner
8           android:id = "@ + id/sp_dropdown"
9           android:layout_width = "match_parent"
10          android:layout_height = "wrap_content"
11          android:spinnerMode = "dropdown" />"
12  </LinearLayout>
```

（3）在 MainAcitivity.java 中，创建适配器对象并对 Spinner 进行设置。

```
1   public class MainActivity extends Activity implements OnItemSe-lectedListener{
2       private String[] starArray;
3       @Override
4       protected void onCreate(Bundle savedInstanceState){
5           super.onCreate(savedInstanceState);
6           setContentView(R.layout.activity_main);
7           starArray = new String[]{"水星","金星","地球","火星",
8                                    "木星","土星"};
9           //创建适配器对象
```

```
10          ArrayAdapter<String> starAdapter1 = new ArrayAdapter<String>
11    (this,R.layout.item_selector,starArray);
12          //设置下拉列表项的布局
13          starAdapter1.setDropDownViewResource(R.layout.item_dropdown);
14          Spinner sp_dropdown =(Spinner) findViewById(R.id.sp_dropdown);
15          sp_dropdown.setPrompt("请选择行星");
16          sp_dropdown.setAdapter(starAdapter1);
17          sp_dropdown.setSelection(0);
18          sp_dropdown.setOnItemSelectedListener(this);      }
19      @Override
20      public void onItemSelected(AdapterView<?>arg0,View arg1,int arg2,
21              long arg3){
22          //TODO Auto-generated method stub
23          Toast.makeText(MainActivity.this,"您选择的是" + starArray[arg2],
24                          Toast.LENGTH_LONG).show();
25      }
26      @Override
27      public void onNothingSelected(AdapterView<?>arg0){
28          //TODO Auto-generated method stub
29      }
    }
```

三、AlertDialog 对话框的使用

AlertDialog 对话框一般以小窗口的形式显示在界面上，用于提示一些重要信息或需要用户交互的内容，其主要组成部分如图 1 – 25 所示。

一般情况下，创建 AlertDialog 对话框包含以下几个步骤。

（1）调用 AlertDialog 的静态内部类 Builder 创建 AlertDialog.Builder 对象。

（2）调用 AlertDialog.Builder 的 setTitle( ) 和 setIcon( ) 方法设置对话框的标题名称和图标。

（3）调用 AlertDialog.Builder 的 setMessage( )、setSingleChoiceItems( ) 等方法设置对话框的提示信息、单选选项等显示内容。

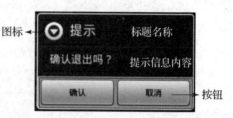

图 1 – 25　对话框的组成

（4）调用 AlertDialog.Builder 的 setPositiveButton( ) 和 setNegativeButton( ) 方法设置对话框的"确认"和"取消"按钮。

（5）调用 AlertDialog.Builder 的 create( ) 方法创建 AlertDialog 对象。

（6）调用 AlertDialog 对象的 show( ) 方法显示对话框。

（7）调用 AlertDialog 对象的 dismiss( ) 方法取消对话框。

（一）普通对话框

普通对话框的提示信息内容一般只显示文本信息。它通过 AlertDialog.Builder 的 setMessage( )方法设置文本信息的具体内容。我们通过下面的案例具体学习普通对话框的创建方法。

【案例1-7】 单击按钮弹出如图1-26所示的对话框,单击"确定"按钮,其文字颜色改为红色。

图1-26 修改文字颜色对话框效果图

1. 案例分析

本案例中对话框需设置图标、标题、内容及两个按钮,在"确定"按钮的单击事件监听中需将按钮上的文字颜色改为红色。

2. 实现步骤

(1) 创建名为"DialogDemo"的程序,并在布局中加入 Button 控件。

(2) 在 MainActivity.java 中,定义 showDialog( )创建对话框,并在按钮的单击事件中调用此方法。

```
1   public class MainActivity extends AppCompatActivity {
2       private Button button;
3       @Override
4       protected void onCreate(Bundle savedInstanceState) {
5           super.onCreate(savedInstanceState);
6           setContentView(R.layout.activity_main);
7           //进行控件初始化
8           init();
9       }
10      private void init() {
11          //获取控件
12          button = findViewById(R.id.bt_dialog1);
13          //添加单击事件监听
14          button.setOnClickListener(new View.OnClickListener() {
15              @Override
16              public void onClick(View view) {
17                  showDialog();
18              }
19          });
20      }
21      //弹出普通提示对话框
22      public void showDialog() {
23          AlertDialog.Builder builder = new AlertDialog.Builder(this);
```

```
24          //设置提示框的标题及图标
25          builder.setTitle("修改文字颜色");
26          builder.setIcon(R.mipmap.ic_launcher);
27          //设置要显示的信息
28          builder.setMessage("你确定让按钮上的文字变成红色吗?");
29          //设置"确定"按钮
30          builder.setPositiveButton("确定",
31                      new DialogInterface.OnClickListener() {
32              @Override
33              public void onClick(DialogInterface dialog,int which) {
34                  button.setTextColor(0xffff0000);
35              }
36          });
37          builder.setNegativeButton("取消",
38                      new DialogInterface.OnClickListener() {
39              @Override
40              public void onClick(DialogInterface dialog,int which) {
41                  dialog.dismiss();   //对话框关闭
42              }
43          });
44          //生成对话框
45          AlertDialog alertDialog = builder.create();
46          //显示对话框
47          alertDialog.show();
48      }
```

上述代码中,第24~44行设置了对话框的具体样式。通过 AlertDialog.Builder 的 setTitle()、setIcon()、setMessage()、setPositiveButton()、setNegativeButton()设置对话框的标题、图标、提示信息、"确定"按钮和"取消"按钮。其中,setPositiveButton()、setNegativeButton()的参数含义相同,第1个参数用来设置按钮的显示内容,第2个参数用来设置按钮的单击事件监听,当不需要监听按钮单击事件时,可以设为"null"。

> **长知识:**
>
> **对话框中的按钮**
> AlertDialog 有3个方法可以设置对话框中的按钮,其中,setPositiveButton()用来设置右边的按钮,setNegativeButton()用来设置左边的按钮,setNeutralButton()用来设置中间的按钮。

(二)单选对话框及多选对话框

单选对话框和多选对话框可以在内容区域显示单选选项列表或多选选项列表。它们的实现是通过调用 AlertDialog.Builder 对象的 setSingleChoiceItems()方法或 setMultiChoiceItems()方法设置的。两个方法具体参数定义如下。

```
1   setSingleChoiceItems(CharSequence[]items,int checkedItem,
                                          OnClickListener listener)
2   setMultiChoiceItems (CharSequence[]items,boolean[]checkedItems,
                                          OnClickListener listener)
```

两个方法的第1个和第3个参数相同，第1个参数 items 表示单选列表或多选列表的选项数据数组；第3个参数 listener 表示列表的事件监听；第2个参数含义相同，都表示默认勾选的选项，setSingleChoiceItems()中 checkedItem 为默认勾选的选项下标，setMultiChoiceItems()中的 checkedItems 为选项是否勾选的 boolean 数组。

【案例1-8】 如图1-27、图1-28所示，单击按钮分别弹出图中所示的单选和多选对话框，并将选择内容显示在下方文本框中。

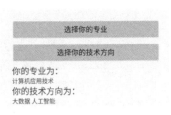

图1-27 按钮及显示界面效果

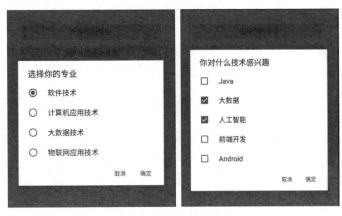

图1-28 单选及多选对话框效果图

1. 案例分析

本案例为"选择你的专业"及"选择你的技术方向"添加单击事件，弹出单选及多选对话框。在对话框的选项列表单击事件的处理方法中可以获得选中的选项 ID，通过 ID 即可获得选项内容。单击对话框的"确定"按钮，将选项内容显示在页面上。

2. 实现步骤

（1）创建名为"ChoiceDialogDemo"的程序，并在布局文件 activity_main.xml 中加入 Button 及 TextView 控件。

（2）在 MainActivity.java 中，定义 singleChoiceDialog() 和 multiChoiceDialog() 方法创建对话框，并在按钮的单击事件中调用方法。

```
1   public class MainActivity extends AppCompatActivity implements
2                                       View.OnClickListener{
3       private Button bt_singlechoice;
4       private Button bt_multichoice;
5       private TextView tv_major,tv_field;
6       private int choiceId;    //单选列表选项下标
7       private String[] major = new String[]{"软件技术","计算机应用技术",
```

```
8                                           "大数据技术","物联网应用技术"};
9         private String[] field = new String[]{"Java","大数据","人工智能",
10                                  "前端开发","Android"};
11        private List<String> list = new ArrayList<String>();
12        @Override        //多选列表选项下标集合
13        protected void onCreate(Bundle savedInstanceState) {
14            super.onCreate(savedInstanceState);
15            setContentView(R.layout.activity_main);
16            init();    //控件初始化
17        }
18        private void init() {
19            bt_singlechoice = findViewById(R.id.bt_dialog3);
20            bt_multichoice = findViewById(R.id.bt_dialog4);
21            tv_major = findViewById(R.id.tv_major);
22            tv_field = findViewById(R.id.tv_field)
23            bt_singlechoice.setOnClickListener(this);
24            bt_multichoice.setOnClickListener(this);
25        }
26        @Override
27        public void onClick(View v) {
28            switch (v.getId()) {
29                case R.id.bt_singlechoice:
30                    singleChoiceDialog ();
31                    break;
32                case R.id.bt_multichoice:
33                    mulitChoiceDialog ();
34                    break;
35            }
36        }
37        //设置单选对话框
38        public void singleChoiceDialog() {
39            AlertDialog.Builder builder = new AlertDialog.Builder(this);
40            builder.setTitle("选择你的专业")
41                    .setSingleChoiceItems(major,0,
42                     new DialogInterface.OnClickListener() {
43                        @Override
44                        public void onClick(DialogInterface dialog,int which) {
45                            choiceId = which;
46                        }
47                    })
48                    .setPositiveButton("确定",
49                     new DialogInterface.OnClickListener() {
50                        @Override
```

```
51                    public void onClick(DialogInterface dialog,int
52                                        which){
53                        tv_major.setText(major[choiceId]);
54                    }
55                })
56                .setNegativeButton("取消",null)
57                .create().show();
58    }
59    //设置多选对话框
60    public voidmulitChoiceDialog(){
61        boolean[] checkedItems ={false,false,false,false,false};
62        AlertDialog.Builder builder = new AlertDialog.Builder(this);
63        builder.setTitle("你对什么技术感兴趣")
64                .setMultiChoiceItems(field,checkedItems,
65                    new DialogInterface.OnMultiChoiceClickListener(){
66                        @Override
67                        public void onClick(DialogInterface dialog,
68                                int which,boolean isChecked){
69                            if (isChecked)
70                                list.add(field[which]);    //将选中项加入集合
71                            else
72                                list.remove(field[which]);  //将未选中项移出集合
73                        }
74                    })
75                .setPositiveButton("确定",
76                    new DialogInterface.OnClickListener(){
77                     @Override
78                     public void onClick(DialogInterface dialog,int which){
79                        StringBuffer info = new StringBuffer();
80                        for(int i =0;i<list.size();i++){
81                            info.append(list.get(i)+" ");}
82                        tv_field.setText(info);
83                     }
84                }).setNegativeButton("取消",null)
85                .create().show();
86    }
87 }
```

第 38~57 行定义了单选对话框，其中，第 42~47 行定义了选项单击事件监听，此处 onClick( ) 方法中的参数 which 代表单击项的下标，将其赋值给 choiceId 变量；第 49~54 行定义了对话框中按钮的单击事件，此处 onClick( ) 方法中的参数 which 的含义与前面不同，它代表对话框中的按钮，在方法中通过前面获取的 choiceId 从选项数组中拿出选中项设置到 TextView 控件上。

第 59～85 行定义了多选对话框，其中，第 64～73 行定义了多选选项单击事件监听，此处 onClick() 方法中的参数 which 同样代表单击项的下标，参数 isChecked 代表是否被选中，通过判断 isChecked，如果被选中，则将选项加入集合 list 中，否则，将选项从集合中移除；在第 75～83 行的"确定"按钮单击事件中，遍历 list 集合取出选中项，并设置到 TextView 控件上。

（三）自定义布局对话框

除了单选、多选对话框，针对不同的界面风格和功能要求，我们也可以根据项目需求创建自定义布局对话框。

【案例 1-9】 单击按钮弹出图 1-29 所示的自定义登录对话框，单击"登录"按钮，将用户名和密码在 Toast 中显示，并关闭对话框。

1. 案例分析

登录对话框布局与系统提供的对话框差距较大，需要自定义对话框布局，并通过 AlertDialog.Builder 对象的 setView() 方法进行设置。对话框中控件的操作与普通界面控件相同，获取控件对象进行所需操作即可。

图 1-29 自定义登录对话框

2. 实现步骤

（1）创建名为"MyDialogDemo"的程序，并在布局文件 activity_main.xml 中加入 Button 控件。

（2）创建 dialog_login.xml 布局文件，编写对话框布局，具体代码如下。

```
1   <?xml version="1.0" encoding="utf-8"?>
2   <LinearLayout xmlns:android="http://schemas.android.com/apk/res/android"
3       android:layout_width="300dp"
4       android:layout_height="wrap_content"
5       android:padding="20dp"
6       android:orientation="vertical">
7       <TextView
8           android:layout_width="wrap_content"
9           android:layout_height="wrap_content"
10          android:layout_gravity="center_horizontal"
11          android:layout_marginTop="10dp"
12          android:text="开启新世界"
13          android:textSize="18sp" />
14      <EditText
15          android:id="@+id/et_username"
16          android:layout_width="match_parent"
17          android:layout_height="wrap_content"
```

```
18          android:layout_marginTop = "40dp"
19          android:background = "@null"
20          android:hint = "请输入用户名"
21          android:textSize = "16sp" />
22      <View
23          android:layout_width = "match_parent"
24          android:layout_height = "1px"
25          android:layout_marginTop = "5dp"
26          android:background = "#D3D3D3" />
27      <EditText
28          android:id = "@+id/et_password"
29          android:layout_width = "match_parent"
30          android:layout_height = "wrap_content"
31          android:layout_marginTop = "30dp"
32          android:background = "@null"
33          android:hint = "请输入密码"
34          android:inputType = "textPassword"
35          android:textSize = "16sp" />
36      <View
37          android:layout_width = "match_parent"
38          android:layout_height = "1px"
39          android:layout_marginTop = "5dp"
40          android:background = "#d3d3d3" />
41      <Button
42          android:id = "@+id/btn_login"
43          android:layout_width = "match_parent"
44          android:layout_height = "40dp"
45          android:layout_marginTop = "35dp"
46          android:background = "#000000"
47          android:padding = "5dp"
48          android:text = "登 录"
49          android:textColor = "#FFFFFF"
50          android:textSize = "16sp" />
51  </LinearLayout>
```

（3）在 MainActivity.java 中，定义 myDialog( ) 方法创建对话框，并在按钮的单击事件中调用方法。

```
1   public class MainActivity extends AppCompatActivity implements
2   View.OnClickListener {
3       private Button bt_login;
4       private EditText et_username,et_password;
5       @Override
6       protected void onCreate(Bundle savedInstanceState) {
7           super.onCreate(savedInstanceState);
```

```java
8          setContentView(R.layout.activity_main);
9          init();//进行控件初始化
10     }
11     private void init() {
12         bt_login = findViewById(R.id.bt_login);
13         bt_login.setOnClickListener(new View.OnClickListener() {
14             @Override
15             public void onClick(View view) {
16                 myDialog();
17             }
18         });
19     }
20     public void myDialog(){
21         //通过对话框布局获取View对象
22         View view = View.inflate(this,R.layout.dialog_login,null);
23         AlertDialog.Builder builder = new AlertDialog.Builder(this);
24         builder.setView(view);       //设置对话框view对象
25         final AlertDialog dialog = builder.create();
26         dialog.show();
27         //通过view对象获取对话框上的控件
28         et_username = view.findViewById(R.id.et_username);
29         et_password = view.findViewById(R.id.et_password);
30         btn_login = view.findViewById(R.id.btn_login);
31         btn_login.setOnClickListener(new View.OnClickListener() {
32             @Override
33             public void onClick(View view) {
34                 //判断输入的用户名和密码是否为空
35                 if(TextUtils.isEmpty(et_username.getText().toString())
36                    ||TextUtils.isEmpty(et_password.getText().toString())) {
37                     Toast.makeText(MainActivity.this,
38                       "用户名或密码不能为空",Toast.LENGTH_SHORT).show();
39                     return;
40                 }
41                 Toast.makeText(MainActivity.this,
42                    "用户名: " + et_username.getText().toString() +
43                    "密码: " + et_password.getText().toString(),
44                    Toast.LENGTH_SHORT).show();
45                 dialog.dismiss();//关闭对话框
46             }
47         });
48     }
49 }
```

在 MainActivity.java 代码的第 20~48 行创建了一个自定义布局对话框,其中,第 22 行调用了 inflate( )方法,该方法根据布局 ID 把这个布局加载成一个 View 对象并返回;第 24 行调用 AlertDialog.Builder 的 setView( )方法将 view 对象设置到对话框上,这样对话框就可以显示自定义布局了;第 28~30 行获取对话框上的控件对象;第 31~47 行给"登录"按钮添加单击事件监听,判断输入的用户名和密码是否为空,如果为空则显示提示信息,不为空则显示用户名和密码,最后关闭对话框。

> **长知识:**
>
> **其他常用对话框**
>
> 除了上述 AlertDialog 的基本用法外,AlertDialog 还派生出多个子类对话框用于实现更为丰富的对话框,比如日期选择对话框 DatePickerDialog、时间选择对话框 TimePickerDialog、进度条对话框 ProgressDialog 等。以下是 DatePickerDialog 和 TimePickerDialog 的构造方法。
>
> • DatePickerDialog(Context context, OnDateSetListener listener, int year, int monthOfYear, int dayOfMonth)
>
> 其中,第 1 个参数为上下文对象,第 2 个参数为日期设置监听器,第 3~5 个参数分别为初始显示的年、月、日。
>
> • TimePickerDialog(Context context, OnTimeSetListener listener, int hourOfDay, int minute, boolean is24HourView)
>
> 其中,第 2 个参数为时间设置监听器,第 3、4 个参数分别为初始显示的小时及分钟;第 5 个参数为是否采用 24 小时制。

任务实施

1. 任务分析

界面布局的实现方式灵活多样,可以直接使用线性布局嵌套,也可以全部采用相对布局,还可以采用线性布局与相对布局相结合的方式实现。这里采用第 3 种方式,整体采用垂直线性布局,界面中的每一行信息采用相对布局,如图 1-30 所示。

修改昵称对话框是一个自定义布局对话框,需要自行实现对话框布局。

2. 实现步骤

(1) 在任务 1 项目的基础上右击,依次选择 New→Activity→Empty Activity,将 Activity Name 设置为"MineActivity",Layout Name 设置为"activity_mine",单击"Finish"按钮,创建"我的信息"页面相关文件。

图 1-30 注册界面效果

(2) 在布局文件 activity_mine.xml 中编写页面布局代码。

【代码 1-5】 activity_mine.xml

```xml
1   <?xml version = "1.0" encoding = "utf-8"?>
2   <LinearLayout xmlns:android = "http://schemas.android.com/apk/res/android"
3       xmlns:tools = "http://schemas.android.com/tools"
4       android:layout_width = "match_parent"
5       android:layout_height = "match_parent"
6       android:background = "@color/gray_tint"
7       android:orientation = "vertical" >
8       <!-- 顶部的实现 -->
9       <RelativeLayout
10          android:layout_width = "match_parent"
11          android:layout_height = "50dp"
12          android:background = "@color/white" >
13          <ImageView
14              android:id = "@+id/mine_back"
15              android:layout_width = "60dp"
16              android:layout_height = "match_parent"
17              android:padding = "12dp"
18              android:src = "@drawable/left_black" />
19          <TextView
20              android:layout_width = "wrap_content"
21              android:layout_height = "match_parent"
22              android:layout_centerInParent = "true"
23              android:gravity = "center"
24              android:text = "我的信息"
25              android:textColor = "@color/black"
26              android:textSize = "22sp"
27              android:textStyle = "bold" />
28      </RelativeLayout>
29
30      <!-- "我的信息"的实现 -->
31      <LinearLayout
32          android:layout_width = "match_parent"
33          android:layout_height = "wrap_content"
34          android:layout_marginTop = "20dp"
35          android:background = "@color/white"
36          android:orientation = "vertical"
37          android:paddingLeft = "10dp"
38          android:paddingRight = "10dp" >
39          <!-- 头像部分省略 -->
40          <!-- 昵称 -->
41          <RelativeLayout
42              android:id = "@+id/user_name"
43              android:layout_width = "match_parent"
44              android:layout_height = "50dp" >
```

```xml
45            <TextView
46                android:layout_width = "wrap_content"
47                android:layout_height = "match_parent"
48                android:gravity = "center"
49                android:text = "昵称"
50                android:textColor = "@color/black"
51                android:textSize = "22sp" />
52            <ImageView
53                android:id = "@+id/user_name_right"
54                android:layout_width = "20dp"
55                android:layout_height = "20dp"
56                android:layout_alignParentEnd = "true"
57                android:layout_centerVertical = "true"
58                android:layout_marginStart = "10dp"
59                android:src = "@drawable/right" />
60            <TextView
61                android:id = "@+id/user_name_tx"
62                android:layout_width = "wrap_content"
63                android:layout_height = "wrap_content"
64                android:layout_centerVertical = "true"
65                android:layout_toStartOf = "@+id/user_name_right"
66                android:text = "小白"
67                android:textSize = "16sp" />
68        </RelativeLayout>
69    <!-- 分割线 -->
70    <View
71            android:layout_width = "match_parent"
72            android:layout_height = "1dp"
73            android:background = "@color/gray" />
74    <!-- 学号部分省略 -->
75    <!-- 电话部分省略 -->
76    <!-- 性别 -->
77    <RelativeLayout
78            android:id = "@+id/user_gender"
79            android:layout_width = "match_parent"
80            android:layout_height = "50dp" >
81            <TextView
82                android:layout_width = "wrap_content"
83                android:layout_height = "match_parent"
84                android:gravity = "center"
85                android:text = "性别"
86                android:textColor = "@color/black"
87                android:textSize = "22sp" />
88            <ImageView
89                android:id = "@+id/user_gender_right"
90                android:layout_width = "20dp"
```

```
91                  android:layout_height = "20dp"
92                  android:layout_alignParentEnd = "true"
93                  android:layout_centerVertical = "true"
94                  android:layout_marginStart = "10dp"
95                  android:src = "@drawable/right" /
96            <TextView
97                  android:id = "@ + id/user_gender_tx"
98                  android:layout_width = "wrap_content"
99                  android:layout_height = "wrap_content"
100                 android:layout_centerVertical = "true"
101                 android:layout_toStartOf = "@ + id/user_gender_right"
102                 android:text = "男"
103                 android:textSize = "16sp" />
104
105        </RelativeLayout>
106        <!-- 分割线 -->
107        <View
108            android:layout_width = "match_parent"
109            android:layout_height = "1dp"
110            android:background = "@color/gray" />
111        <!-- 生日 -->
112        <RelativeLayout
113            android:id = "@ + id/user_birthday"
114            android:layout_width = "match_parent"
115            android:layout_height = "50dp" >
116            <TextView
117                  android:layout_width = "wrap_content"
118                  android:layout_height = "match_parent"
119                  android:gravity = "center"
120                  android:text = "生日"
121                  android:textColor = "@color/black"
122                  android:textSize = "22sp" />
123            <ImageView
124                  android:id = "@ + id/user_birthday_right"
125                  android:layout_width = "20dp"
126                  android:layout_height = "20dp"
127                  android:layout_alignParentEnd = "true"
128                  android:layout_centerVertical = "true"
129                  android:layout_marginStart = "10dp"
130                  android:src = "@drawable/right" />
131            <TextView
132                  android:id = "@ + id/user_birthday_tx"
133                  android:layout_width = "wrap_content"
134                  android:layout_height = "wrap_content"
135                  android:layout_centerVertical = "true"
```

```xml
                    android:layout_toStartOf = "@ + id/user_birthday_right"
                    android:text = "未设置"
                    android:textSize = "16sp" />
        </RelativeLayout>
        <!-- 分割线 -->
        <View
            android:layout_width = "match_parent"
            android:layout_height = "1dp"
            android:background = "@color/gray" />

        <!-- 地区 -->
        <RelativeLayout
            android:layout_width = "match_parent"
            android:layout_height = "50dp" >
            <TextView
                android:layout_width = "wrap_content"
                android:layout_height = "match_parent"
                android:gravity = "center"
                android:text = "学院"
                android:textColor = "@color/black"
                android:textSize = "22sp" />
            <Spinner
                android:id = "@ + id/college_sp"
                android:layout_width = "wrap_content"
                android:layout_height = "wrap_content"
                android:layout_centerVertical = "true"
                android:layout_alignParentRight = "true"
                android:spinnerMode = "dialog"
                android:textSize = "16sp" />
        </RelativeLayout>

    </LinearLayout>

    <!-- 工作状态 -->
    <RelativeLayout
        android:id = "@ + id/rl_status"
        android:layout_width = "match_parent"
        android:layout_height = "50dp"
        android:layout_marginTop = "20dp"
        android:background = "@color/white"
        android:paddingLeft = "10dp"
        android:paddingRight = "10dp" >
        <TextView
            android:layout_width = "wrap_content"
            android:layout_height = "match_parent"
```

```
181                 android:gravity = "center"
182                 android:text = "学习状态"
183                 android:textColor = "@color/black"
184                 android:textSize = "22sp" />
185             <ImageView
186                 android:id = "@+id/status_right"
187                 android:layout_width = "20dp"
188                 android:layout_height = "20dp"
189                 android:layout_alignParentEnd = "true"
190                 android:layout_centerVertical = "true"
191                 android:layout_marginStart = "10dp"
192                 android:src = "@drawable/right" />
193             <TextView
194                 android:id = "@+id/tv_status"
195                 android:layout_width = "wrap_content"
196                 android:layout_height = "wrap_content"
197                 android:layout_centerVertical = "true"
198                 android:layout_toStartOf = "@+id/status_right"
199                 android:text = "放假中"
200                 android:textSize = "16sp" />
201             <ImageView
202                 android:id = "@+id/iv_status"
203                 android:layout_centerVertical = "true"
204                 android:layout_width = "wrap_content"
205                 android:layout_height = "30dp"
206                 android:layout_toLeftOf = "@+id/tv_status"
207                 android:src = "@drawable/vacation" />
208         </RelativeLayout>
209 </LinearLayout>
```

（3）在 MineActivity.java 中编写页面功能逻辑代码。任务中需要3种对话框,创建3个方法 usenameChangeDialog( )、genderChoiceDialog( )、birthdayDialog( ),分别在相关信息被单击时调用。

【代码1-6】 MineActivity.java

```
1  public class MineActivity extends AppCompatActivity implements
2                                         View.OnClickListener{
3      private RelativeLayout user_name;
4      private TextView user_name_tx;
5      private RelativeLayout user_gender;
6      private TextView user_gender_tx;
7      private RelativeLayout user_birthday;
8      private TextView user_birthday_tx;
9      private Spinner college_sp;
10     private String[] college = new String[]{"智慧学院","经贸学院",
```

```
11                              "机电学院","旅游学院","交通学院"};
12      @Override
13      protected void onCreate(Bundle savedInstanceState) {
14          super.onCreate(savedInstanceState);
15          setContentView(R.layout.activity_mine);
16          //初始化视图
17          initView();
18      }
19
20      private void initView() {
21          //找到对应的控件
22          user_name = findViewById(R.id.user_name);
23          user_name_tx = findViewById(R.id.user_name_tx);
24          user_gender = findViewById(R.id.user_gender);
25          user_gender_tx = findViewById(R.id.user_gender_tx);
26          user_birthday = findViewById(R.id.user_birthday);
27          user_birthday_tx = findViewById(R.id.user_birthday_tx);
28          college_sp = findViewById(R.id.college_sp);
29          //学院控件初始化
30          ArrayAdapter adapter = new ArrayAdapter(this
31                  ,android.R.layout.simple_list_item_1,college);
32          college_sp.setAdapter(adapter);
33          college_sp.setPrompt("请选择学院:");
34          //设置单击事件监听
35          user_name.setOnClickListener(this);
36          user_gender.setOnClickListener(this);
37          user_birthday.setOnClickListener(this);
38      }
39      @Override
40      public void onClick(View v) {
41          switch (v.getId()) {
42              case R.id.user_name:
43                  usenameChangeDialog();   //调用昵称修改对话框方法
44                  break;
45              case R.id.user_gender:
46                  genderChoiceDialog();    //调用性别选择对话框方法
47                  break;
48              case R.id.user_birthday:
49                  birthdayDialog();        //调用出生日期选择对话框方法
50                  break;
51              default:
52                  break;
53          }
54      }
55  //昵称修改对话框
```

```java
56  public void usenameChangeDialog(){
57      final View layout = View.inflate(this,R.layout.dialog_user_name,null);
58      AlertDialog.Builder builder = new AlertDialog.Builder(this);
59      builder.setTitle("修改昵称")
60              .setView(layout)
61              .setPositiveButton("确定",new DialogInterface.OnClick-Listener()
62              { @Override
63                  public void onClick(DialogInterface dialog,int id) {
64                      EditText update_user_name =
65                              layout.findViewById(R.id.update_user_name);
66                      String newName =
67                              update_user_name.getText().toString().trim();
68                      //设置控件中的数据
69                      user_name_tx.setText(newName);
70                  }
71              })
72              .setNegativeButton("取消",new DialogInterface.OnClick-Listener()
73              { @Override
74                  public void onClick(DialogInterface dialog,int id) {
75                  }
76              });
77      AlertDialog dialog = builder.create();
78      dialog.show();
79  }
80  //性别选择单选对话框
81  public void genderChoiceDialog(){
82      final String[] genders = new String[]{"男","女"};
83      AlertDialog.Builder builder1 = new AlertDialog.Builder(this);
84      builder1.setTitle("性别选择")
85              .setSingleChoiceItems(genders,0,
86                  new DialogInterface.OnClickListener() {
87                      @Override
88                      public void onClick(DialogInterface dialog,int which) {
89                          user_gender_tx.setText(genders[which]);
90                          dialog.dismiss();
91                      }
92                  });
93      AlertDialog dialog1 = builder1.create();
94      dialog1.show();
95  }
96  //选择日期对话框
97  public void birthdayDialog() {
98      Calendar ca = Calendar.getInstance();
99      int  mYear = ca.get(Calendar.YEAR);
100     int  mMonth = ca.get(Calendar.MONTH);
```

```
101        int  mDay = ca.get(Calendar.DAY_OF_MONTH);
102        DatePickerDialog dialog1 = new DatePickerDialog(this,
103            new DatePickerDialog.OnDateSetListener() {
104            //日期选择器上的月份是从 0 开始的
105            @Override
106            public void onDateSet(DatePicker view,int year,
107                                 int monthOfYear,int dayOfMonth)
108            {
109                user_birthday_tx.setText(year + "年" +
110                (monthOfYear +1) + "月" + dayOfMonth + "日");}
111            },mYear,mMonth,mDay);
112        //显示时间的对话框
113        dialog1.show();
114     }
115 }
```

任务反思

实现并运行本任务，将在代码编写及程序调试过程中出现的异常信息、产生原因及解决方法记录在下方。

问题1：_____

产生原因：_____

_____

解决方法：_____

_____

问题2：_____

产生原因：_____

_____

解决方法：_____

任务总结及巩固

师兄：小白，这部分我们学习了新的界面布局及控件，界面效果和能实现的功能也丰富起来了。来，你再来总结一下。

小白：嗯，我来试试。
- 相对布局找基准，基准一定先定义。
- 十五属性分四组，找完父亲找兄弟。

- 控件各自有特点，关系属性需弄清。
- 下拉列表适配器，两者合作展数据。
- 莫把事件忘处理，选项选中才有效。

师兄：总结得越来越到位了，这些一定要在不断的代码实践中，加深理解，熟练应用。上面的任务还有一个问题，原来我们是让 MainActivity 也就是登录页面作为首页面显示的，你是怎么让同一个项目中"我的信息"页面运行在模拟器上的呢？

一、基础巩固

1. 在相对布局中，如果想要将一个 Button 控件放在另一个 Button 控件的右侧，可以使用的属性是（　　）。

   A. android：layout_toRightOf　　　　B. android：layout_below

   C. android：layout_alignParentRight　D. android：layout_toLeftOf

2. 在相对布局中，如果想要将一个控件放在布局的中间，可以使用的属性是（　　）。

   A. android：layout_centerInParent

   B. android：layout_centerHorizontal

   C. android：gravity

   D. android：layout_gravity

3. 下面关于 AlertDialog 的描述错误的是（　　）。

   A. 对话框的显示需要调用 show( ) 方法

   B. setPositiveButton( ) 方法用来设置"确定"按钮

   C. 需要创建 AlertDialog.Builder 对象进行对话框的设置

   D. 使用 new 关键字创建 AlertDialog 实例

4. 在设置 Spinner 控件的列表项时，除了可以使用 android：entries 属性外，还可以通过（　　）来加载。

   A. Listener　　　　　　　　　　　B. Adapter

   C. Manager　　　　　　　　　　　D. Activity

5. 在使用 Spinner 时，说法错误的是（　　）。

   A. 需要在布局中，加入 Spinner 控件

   B. Spinner 只有下拉列表一种形式

   C. Adapter 将列表项数据放入列表项布局中

   D. 通过 setOnItemSelectedListener( ) 方法设置列表项选中事件监听器

二、技术实践

利用本任务中学习的内容完成如图 1－3 所示的"学习状态"页面。

任务3　界面间跳转的实现

任务描述

在任务2的基础上，进一步完成页面跳转及数据传递功能，如图1-31所示。

单击"我的信息"界面中的"学习状态"选项，跳转至"学习状态"页面，将当前学习状态在"学习状态"页面上方显示，并将新选择的学习状态返回到"我的信息"界面中显示。

图1-31　任务3效果图

任务学习目标

通过本任务需达到以下目标：
➢ 理解Activity的生命周期及相关方法的作用。
➢ 能够灵活运用Intent实现Activity间的消息传递。

技术储备

一、认识Activity

认识Activity

从创建HelloWorld项目起，Activity就已经进入我们的视线。Activity代表活动，是Android四大组件之一。一个Activity就是一个可以与用户进行交互的界面，在一个应用程序中可以包含多个Activity，当然也有个别应用程序一个Activity也没有。

（一）创建Activity

每个应用程序在创建时都会自动生成一个MainActivity.java，并伴随有布局文件activity_main.xml。布局文件中包含了Activity的布局、控件等，Java源文件中完成了Activity中的功能逻辑。同时，在 AndroidManifest.xml 有对 MainActivity 的声明。当我们需要新建一个 Activity 时同样需要完成以下 3 个步骤。

（1）在 res/layout 目录下创建布局文件。

（2）在 src 目录下创建 XXXActivity 类，继承 Activity 或其子类，并实现 onCreate（）方法，用于初始化 Activity 中必需的组件，并需要调用 setContentView（）方法，以设置 Activity 用户界面的布局。

（3）在 AndroidManifest.xml 中声明 Activity。

> **注意**：在 Android Studio 中可以通过上述 3 个步骤，实现一个 Activity；也可以直接在 module 上右击，依次选择 New→Activity，选择需要的 Activity 类型，将上述 3 个步骤三合一。

（二）结束 Activity

通过调用 Activity 的 finish（）方法可以结束一个 Activity。但大多数情况下，我们不会调用方法显式结束 Activity，而是交由 Android 系统对 Activity 进行管理。

（三）Activity 生命周期

一个 Activity 从创建到结束的全过程称为 Activity 的生命周期。Activity 的生命周期由系统统一管理，Android 系统通过活动栈来管理 Activity，每当我们启动了一个新的 Activity，它会在活动栈中入栈，并处于栈顶的位置。而每当我们按下 Backspace 键或调用 finish（）方法去销毁一个 Activity 时，处于栈顶的 Activity 会出栈，这时前一个入栈的 Activity 就会重新处于栈顶的位置。系统总是会显示处于栈顶的 Activity 给用户，如图 1-32 所示。

1. Activity 的状态

在 Activity 的生命周期中，包含以下 4 种状态。

图 1-32　Android 中的任务栈

- Active/Runing：一个新 Activity 启动入栈后，它在屏幕最前端，处于栈的最顶端，此时它处于可见并可与用户交互的激活状态。
- Paused：当 Activity 被另一个透明的或者 Dialog 样式的 Activity 覆盖时，处于 Paused 状态。此时它依然与窗口管理器保持连接，仍然可见，但它已经失去了焦点，不可与用户交互。
- Stopped：当 Activity 被另外一个 Activity 覆盖、失去焦点并不可见时，处于 Stopped 状态。

- Killed：Activity 被系统杀死回收或者没有被启动时，处于 Killed 状态。

当一个 Activity 实例被创建、销毁或者启动另外一个 Activity 时，它在这 4 种状态之间进行转换，这种转换的发生依赖于用户程序的动作。Activity 在不同状态间转换的时机和条件如图 1-33 所示。

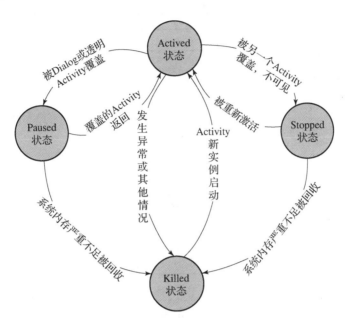

图 1-33　Activity 的状态转换

2. Activity 的生命周期方法

Activity 类中定义了 7 个生命周期方法，在 Activity 状态转换过程中，这些方法将会被调用，并发挥其特定作用。生命周期方法及其功能描述见表 1-7。Activity 生命周期中状态变化及方法回调如图 1-34 所示。

表 1-7　Activity 的生命周期方法及其功能描述

方法名	功能描述
onCreate( )	创建活动，完成对活动的初始化操作，在活动第一次被创建的时候调用
onStart( )	开始活动，使活动显示在屏幕上，在活动由不可见变为可见时调用
onResume( )	恢复活动，在活动准备好和用户进行交互时调用
onPause( )	暂停活动，让活动在屏幕上的动作暂停，在系统准备去启动或者恢复另一个活动时调用
onStop( )	停止活动，把活动从屏幕上撤下来，在活动完全不可见时调用
onDestroy( )	销毁活动，把活动从内存中清除掉，在活动被销毁之前调用
onRestart( )	重启活动，重新加载内存中的活动数据，在活动由停止状态变为运行状态之前调用

【案例 1-10】　验证启动 App、竖屏与横屏切换、按下 Home 键并返回 App、单击回退键退出程序 4 种情况下，Activity 的生命周期方法的调用情况。

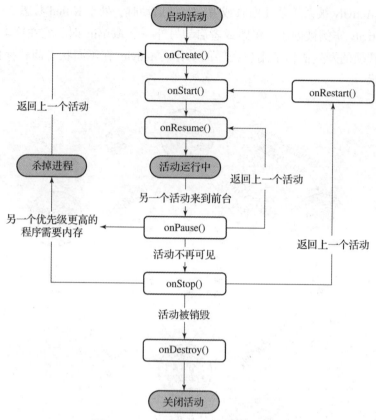

图 1-34 Activity 生命周期方法调用

1. 案例分析

本案例要验证在 Activity 状态发生变化过程中，生命周期方法的调用情况，可以通过在生命周期方法中加入日志输出语句，通过日志输出观察哪些生命周期方法被调用。

2. 实现步骤

（1）创建名为"LifeCycleDemo"的程序。

（2）在 MainActivity.java 中添加生命周期方法。

在 MainActivity 类中空白处右击，依次选择 Generate→Override Methods，选择生命周期方法。

（3）在生命周期方法中，加入 Log.i() 方法，运行时在日志中输出生命周期方法的调用信息。

具体代码如下。

```
1  public class MainActivity extends AppCompatActivity {
2      @Override
3      protected void onCreate(Bundle savedInstanceState) {
4          super.onCreate(savedInstanceState);
5          setContentView(R.layout.activity_main);
6          Log.i("LifeDemo","onCreate 被调用");
7      }
```

```java
8   @Override
9   protected void onPause() {
10      super.onPause();
11      Log.i("LifeDemo","onPause 被调用");
12  }
13  @Override
14  protected void onResume() {
15      super.onResume();
16      Log.i("LifeDemo","onResume 被调用");
17  }
18  @Override
19  protected void onStart() {
20      super.onStart();
21      Log.i("LifeDemo","onStart 被调用");
22  }
23  @Override
24  protected void onStop() {
25      super.onStop();
26      Log.i("LifeDemo","onStop 被调用");
27  }
28  @Override
29  protected void onDestroy() {
30      super.onDestroy();
31      Log.i("LifeDemo","onDestroy 被调用");
32  }
33  @Override
34  protected void onRestart() {
35      super.onRestart();
36      Log.i("LifeDemo","onRestart 被调用");
37  }
38  }
```

启动 App，进入 MainActivity 页面。MainActivity 进入活动状态，在 Logcat 中，打印出了进入活动状态过程中，回调方法的执行情况，如图 1-35 所示。

图 1-35 回调方法的执行情况

从执行情况来看，onCreate()、onStart()、onResume()方法先后被调用，Activity 进入可见、可交互的活动状态。

模拟器从竖屏切换至横屏。从图 1-36 所示的 Logcat 日志打印信息看，在竖屏切换至横屏时，onPause()、onStop()、onDestroy()3 个方法被依次调用，活动被销毁；接着，onCreate()、onStart()、onResume()方法依次调用，重新启动进入活动状态。

图 1-36　竖屏切换至横屏时的执行情况

按下 Home 键并返回 App。单击 Home 键回到手机屏幕，再次单击 App 应用图标打开 MainActivity。此时，MainActivity 并没有被销毁而是不可见，进入了停止状态，再次打开后又再次进入活动状态，方法调用执行情况如图 1-37 所示。

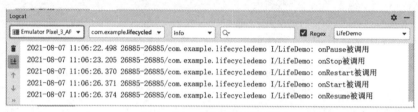

图 1-37　方法被调用情况

从运行结果看，onPause( )、onStop( ) 方法首先被调用，MainActivity 进入停止状态，再次启动活动，onReStart( )、onStart( )、onResume( ) 依次被调用来到前台。

单击回退键，退出程序。单击模拟器上的回退键，可以看到 MainActivity 被销毁，程序退出，方法被调用情况如图 1-38 所示。

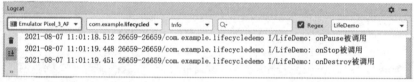

图 1-38　退出程序时方法被调用情况

从打印的日志信息可以看到，onPause( )、onStop( )、onDestroy( ) 方法依次被调用，Activity 进入销毁状态，并被清理出内存。

> **长知识：**
>
> （1）如何设置 Activity 在屏幕转换时不被销毁？
>
> 横竖屏转换时活动被销毁有时会给应用带来麻烦，比如用户正在应用中填写个人信息，不小心转换手机方向，页面被销毁，又重新启动，用户填写的信息也丢失了，这是一种不愉快的用户体验。
>
> 可以在 AndroidManifest.xml 相应的 Activity 节点中加入如下属性设置。
>
> ```
> <activity android:name=".MainActivity"
> android:configChanges="orientation|keyboardHidden|screenSize">
> ```

(2) 如何设置屏幕方向不能变化？

可以在 AndroidManifest.xml 相应的 Activity 节点中加入如下属性设置。

android:screenOrientation = "portrait"   //竖屏

android:screenOrientation = "landscape"   //横屏

二、认识 Intent

认识 Intent

Intent 表示意图，它用来进行组件间的消息传递，也就是说明想让哪个组件干什么。它可以在 Activity 之间进行消息传递，比如我们有两个 Activity，需要通过单击第一个 Activity 上的按钮启动第二个 Activity，这就需要 Intent 来传递这个消息。同时，Activity 与 Service（服务）及 Broadcast（广播）间的消息传递也靠 Intent。

（一）Intent 的主要任务

为了做好组件间的消息传递，Intent 需要完成 3 个主要任务。

(1) 明确消息从哪里来、到哪里去、用什么方式传递。

(2) 携带好本次消息传递的数据内容，数据在发起方装包，在接收方解包。

(3) 如果发起方需要进一步判断接收方对数据处理的结果或对接收方传回的数据进行处理，Intent 要负责让接收方传回应答数据。

（二）Intent 的组成

为了能够完成以上任务，Intent 需要能够携带足够的信息，以下是 Intent 的 6 个主要组成部分。

1. Component（组件）

用于指定 Intent 的来源及目的组件。通过 setClass( )、setComponent( )、setClassName( ) 设置，通过 getComponent( ) 获取。

2. Action（动作）

用于指定 Intent 要完成的动作。其值为一个字符串常量，这个值可以是用户自定义的常量，如果是自定义常量需要加上应用的包名作为前缀，也可以是 Intent 中预定义的常量。常用的 Intent 预定义常量见表 1-8。

表 1-8 常用的 Intent 预定义常量

常量名	常量值	说明
ACTION_MAIN	android.intent.actiong.MAIN	App 启动时的入口
ACTION_VIEW	android.intent.actiong.VIEW	显示数据给用户，如打开浏览器
ACTION_CALL	android.intent.actiong.CALL	直接拨打电话
ACTION_DIAL	android.intent.actiong.DIAL	进入拨号界面
ACTION_SENDTO	android.intent.actiong.SENDTO	发短信

使用 setAction( ) 和 getAction( ) 来设置和读取 Action 属性。

3. Data（数据规格）

动作的 URI 类型，对于不同的 Action 有不同的数据规格。比如，Action 是"ACTION_CALL"时，数据是要拨打的电话号码，数据格式为"tel：电话号码"；如果要通过浏览器打开百度页面，则 Action 需要为"ACTION_VIEW"，数据是百度的网址，具体格式为"http://www.baidu.com"。

通过 setData( ) 方法可以指定数据为 URI，getData( ) 方法可以读取 URI。

4. Category（种类）

Category 能够处理这个 Intent 的组件种类，其值为一个字符串。在 Intent 类中定义了一些 Category 常量，见表 1-9。

表 1-9 Category 常量

常量名	说明
CATEGORY_DEFAULT	Android 系统中默认的执行方式，所有组件都能处理此 Intent
CATEGORY_HOME	表示设备启动（登录屏幕）时显示的第一个 Activity
CATEGORY_LAUNCHER	表示该 Activity 作为应用程序的启动项
CATEGORY_BROWSABLE	表示该 Activity 能够被浏览器安全调用

与 Category 相对应的方法有添加种类 addCategory( )、移除种类 removeCategory( ) 和获取所有种类 getCategories( )。

5. Extras（额外数据）

Extras 用于指定 Intent 需要传递的额外数据。比如通过单击 Activity1 中的按钮启动 Activity2，并将 Activity1 中输入的姓名、年龄、性别等信息传递给 Acivity2 显示，就需要将这些附加信息作为 Extras 传递。

6. Flags（标志位）

各种类型的 Flag，很多是用来指定 Android 系统如何启动 activity，以及启动了 activity 后如何对待它。所有这些都定义在 Intent 类中。

（三）Intent 的两种方式

Intent 消息传递方式有两种，一种是显式 Intent，另一种是隐式 Intent。

Intent 的两种方式

1. 显式 Intent

（1）特点。

直接指定消息从哪个组件发出，要到哪个组件去。

（2）用法。

通过 Intent 构造方法指定，具体代码如下。

```
1  Intent intent = new Intent(MainActivity.this,SecondActivity.class);
2  startActivity(intent);              //启动 Activity
```

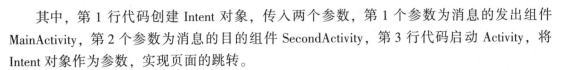

其中，第 1 行代码创建 Intent 对象，传入两个参数，第 1 个参数为消息的发出组件 MainActivity，第 2 个参数为消息的目的组件 SecondActivity，第 3 行代码启动 Activity，将 Intent 对象作为参数，实现页面的跳转。

2. 隐式 Intent

（1）特点。

不明确指定消息的来源和去向，而是通过设置"暗号"，由系统自动进行"暗号"匹配，确定消息的目的组件。通常当我们不希望直接暴露源码的类名或者不知道目的地类名时，会使用隐式 Intent。

（2）用法。

在使用隐式 Intent 时，需要给 Intent 指定 Action、Data 等属性，这些属性就是系统用来确定目的组件的"暗号"。

例如，当我们想要调用浏览器打开百度页面，我们并不知道浏览器页面的类名，无法使用显式 Intent，就可以通过隐式 Intent 完成跳转，具体可以按照如下代码设置 Intent。

```
1   Intent intent = new Intent();              //创建 Intent 对象
2   intent.setAction(Intent.ACTION_VIEW);      //设置 action 为 Intent.ACTION_VIEW
3   Uri uri = Uri.parse("http://www.baidu.com");   //将百度网址转为 URI 对象
4   intent.setData(uri);                       //设置数据
5   startActivity(intent);                     //跳转至浏览器页面
```

上述代码中，第 2 行代码调用 setAction()方法为 Intent 对象设置 action 为"Intent.ACTION_VIEW"，这是 Android 中的预定义常量，具体的字符串值为"android.intent.action.VIEW"；第 3 行代码将百度的网址"http://www.baidu.com"转换为 URI 对象；第 4 行代码调用 intent 对象的 setData()方法设置数据；第 5 行代码调用 startActivity()方法跳转到浏览器页面，并打开百度页面。

在这个过程中，Android 系统会默认为 Intent 添加 category 为"android.intent.category.DEFAULT"，并根据 Intent 的 action、data、category 设置与每一个系统中组件的 IntentFilter（过滤器）进行匹配，这 3 个属性都匹配成功时，就会启动相应的组件。也就是说，在使用隐式 Intent 时，对 Intent 设置的 action 等属性，需要跟目的组件的 IntentFilter 中设置的 action 等属性一致才可以。那 IntentFilter 定义在哪里呢？

Intent 过滤器通常在 AndroidManifest.xml 中进行设置，下面是 MainActivity 的声明代码，其中 < intent – filter > </intent – filter > 中的部分就是 intent 过滤器。

```
1   <activity android:name=".MainActivity">
2       <intent-filter>
3           <action android:name="android.intent.action.MAIN" />
4           <category android:name="android.intent.category.LAUNCHER" />
5       </intent-filter>
6   </activity>
```

此过滤器中设置了 action 和 category 两个过滤条件，其中，android.intent.action.MAIN 表示 App 启动时的入口，android.intent.category.LAUNCHER 表示 App 启动时调用。

如果在我们自己的应用程序中,希望通过隐式 Intent 实现页面间的跳转,就需要为每个组件设置 IntentFilter,这样系统就能够过滤出符合条件的组件并按优先顺序调用。

> **注意**:用户自定义的 Activity 如果希望能够被隐式 Intent 启动,需要在声明时设置 Intent 过滤器,过滤器中需要设置 action,作为系统匹配的动作;同时,category 的设置至少要包含一个"android.intent.category.DEFAULT"。

【案例 1-11】 在任务 1 和任务 2 基础上,实现当在登录页面中输入用户名和密码正确的时候,跳转至"我的信息"页面,如图 1-39 所示。

图 1-39 登录页面跳转效果图

1. 案例分析

使用隐式 Intent 实现跳转,需要为"我的信息"页面 MineActivity 设置 IntentFilter,并在登录页面 MainActivity.java 中创建 Intent 对象,实现页面跳转。

2. 实现步骤及具体代码

(1) 在 AndroidManifest.xml 文件中,为"我的信息"页面添加 IntentFilter,即在 <activity> 节点中添加 <intent-filter>,添加代码如【代码 1-7】所示。

(2) 在 MainActivity.java 中,修改逻辑代码。当用户名和密码正确时,创建 Intent 对象,设置 Action 值为"com.example.learntopass.MINEACTIVITY",并调用 startActivity()方法启动页面,添加代码如【代码 1-8】所示。

【代码 1-7】 AndroidManifest.xml

```
1  <activity android:name=".MineActivity" >
2      <intent-filter>
3          <action android:name="com.example.learntopass.MINEACTIVITY" />
4          <category android:name="android.intent.category.DEFAULT" />
5      </intent-filter>
6  </activity>
```

【代码1-8】 MainActivity.java

```java
1  public class MainActivity extends AppCompatActivity{
2      private Button btn_login;
3      private EditText et_phonenum,et_password;
4      private String phonenum,password;
5      @Override
6      protected void onCreate(Bundle savedInstanceState) {
7          super.onCreate(savedInstanceState);
8          setContentView(R.layout.activity_main);
9          init();
10         btn_login.setOnClickListener(new View.OnClickListener() {
11             @Override
12             public void onClick(View arg0) {
13                 phonenum = et_phonenum.getText().toString();
14                 password = et_password.getText().toString();
15                 if(phonenum = = null || "".equals(phonenum)){
16                     Toast.makeText(MainActivity.this,"请输入手机号码",
17                             Toast.LENGTH_LONG).show();
18                 }else if(password = = null || "".equals(password)){
19                     Toast.makeText(MainActivity.this,"请输入密码",
20                             Toast.LENGTH_LONG).show();
21                 }else if(phonenum.equals("12345678910") &&
22                                    password.equals("123456")){
23                     Intent intent =
24                       new Intent("com.example.learntopass.MINEACTIVITY ");
25                     startActivity(intent);
26                 }else{
27                   Toast.makeText(MainActivity.this,"手机号码或密码错误",
28                           Toast.LENGTH_LONG).show();
29                 }
30             }
31         });
32     }
33     public void init() {
34         btn_login = findViewById(R.id.btn_login);
35         et_phonenum = findViewById(R.id.et_phonenum);
36         et_password = findViewById(R.id.et_password);
37     }
38 }
```

在【代码1-7】AndroidManifest.xml 中，第2~5行设置了 IntentFilter，其中包含 action 值，一般采用包名+类名的形式；category 值至少要设置一个，其值为"android.intent.category.DEFAULT"。

在【代码1-8】MainActivity.java 中，第23、24行定义了隐式 Intent，并通过构造方法设置 action 为 "com.example.learntopass.MINEACTIVITY"，这里注意，此处的 action 应与

AndroidManifest.xml 中 MineActivity 的 IntentFilter 中的 action 一致。第 25 行调用 startActivity() 启动页面。

需要注意的是，每一个 Intent 中只能指定一个 action，但可以指定多个 category。目前，MainActivity.java 中的 intent 只有一个默认的 category，其值就是"android.intent.category.DEFAULT"。我们再通过 addCategory() 方法增加一个 category。

```
1  Intent intent = new Intent("com.example.learntopass.MINEACTIVITY");
2  intent.addCategory("com.example.learntopass.MY_CATEGORY");
3  startActivity(intent);
```

运行程序，输入正确的用户名和密码，单击"登录"按钮，程序发生异常，退出。我们可以在 Run 界面中查看错误信息，如图 1-40 所示。

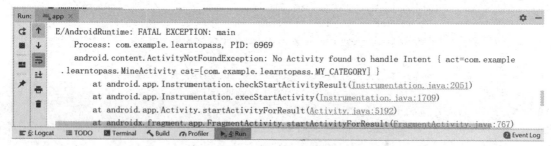

图 1-40 错误信息

错误信息中提示，没有找到响应 Intent 的 Activity。这是由于在 Intent 中设置了两个 category，而 MineActivity 的 IntentFilter 中只有一个 category，系统在匹配的时候不成功。我们可以在 IntentFilter 中再添加一个 category 的声明，上面的问题就迎刃而解了。

```
1  <activity android:name=".MineActivity">
2      <intent-filter>
3          <action android:name="com.example.learntopass.MineActivity" />
4          <category android:name="android.intent.category.DEFAULT" />
5          <category android:name="com.example.learntopass.MY_CATEGORY" />
6      </intent-filter>
7  </activity>
```

三、Activity 间的数据传递

通过 Intent 可以实现 Activity 之间的跳转，下面讨论前面提到的在跳转过程中需要数据传递的应用场景：通过单击 FirstActivity 中的按钮启动 SecondActivity，并将 FirstActivity 中输入的姓名、年龄、性别等信息传递给 SecondActivity 显示。

1. 向下一个 Activity 传递数据

Intent 中重载了多个 putExtra() 方法用于传递不同类型的数据，我们在 FirstActivity 中将数据装入 Intent，在 SecondActivity 中从 Intent 中取出数据，具体代码如下。

向下一个 Activity 传递数据

（1）FirstActivity 中装入数据。

```
1  Intent intent = new Intent(FirstActivity.this,SecondActivity.class);
2  intent.putExtra("name","张三");
3  intent.putExtra("age",19);
4  intent.putExtra("sex","男");
5  startActivity(intent);
```

（2）SecondActivity 中取出数据。

```
1  Intent intent = getIntent();
2  String name = intent.getStringExtra("name");
3  int age = intent.getIntExtra("age",18);
4  String sex = intent.getStringExtra("sex");
```

前面我们将数据一个一个装入 Intent，这就好比我们要邮寄 3 样东西，快递员将它们直接放进了快递车，而实际中我们通常会将 3 样东西装进一个包裹中再交由快递员邮寄。在 Intent 传递数据时，可以通过 Bundle 类实现对数据的打包处理。我们先将数据存入 Bundle 对象中，再将 Bundle 对象交给 Intent，具体代码如下。

（3）FirstActivity 中装入打包数据。

```
1  Intent intent = new Intent(FirstActivity.this,SecondActivity.class);
2  Bundle bundle = new Bundle();              //创建 Bundle 对象
3  bundle.putString("name","张三");           //将数据存入 Bundle 对象
4  bundle.putInt("age",19);
5  bundle.putString("sex","男");
6  intent.putExtras(bundle);                   //将 Bundle 对象交给 intent
7  startActivity(intent);
```

（4）SecondActivity 中取出打包数据。

```
1  Intent intent = getIntent();                      //获取 Intent 对象
2  Bundle bundle = intent.getExtras();               //取出 Bundle 对象
3  String name = bundle.getString("name","");        //取出数据
4  int age = bundle.getInt("age",18);
5  String sex = bundle.getString("sex","");
```

2. 向上一个 Activity 返回数据

在某些场景中，我们还需要让 SecondActivity 对传来数据的处理结果或者处理后的数据返回给 FirstActivity，下面我们分 4 步完成这个过程。

向上一个
Activity
返回数据

（1）FirstActivity 中将数据及信息放入 Intent，调用 startActivityForResult（Intent intent，int requestCode）。与 startActivity（）方法仅用来启动 Activity 不同，此方法表示启动新的 Activity 并需要返回结果数据，其中，第 1 个参数同样是 intent 对象，第 2 个参数是请求码，用来唯一标识本次请求。

```
1   Intent intent = new Intent(FirstActivity.this,SecondActivity.class);
2   Bundle bundle = new Bundle();
3   bundle.putString("name","张三");
4   bundle.putInt("age",19);
5   bundle.putString("sex","男");
6   intent.putExtras(bundle);
7   startActivityForResult(intent,0);
```

（2）SecondActivity 接收 Intent，并进行相应处理，此处代码同前。

（3）SecondActivity 将需要返回的数据放入 Intent 中，并调用 setResult（int ResultCode, Intent intent）方法返回信息。其中，第 1 个参数为结果码，代表返回的状态（比如成功或失败），通常可以用常量 Activity. RESULT_OK 和 Activity. RESULT_CANCLE；第 2 个参数为设置返回数据的 intent 的对象。

```
1   Intent intent = new Intent();
2   Bundle bundle = new Bundle();
3   bundle.putString("info","已接收到数据");
4   intent.putExtras(bundle);
5   setResult(Activity.RESULT_OK,intent);
```

（4）在 FirstActivity 中重写 onActivityResult( )，对返回结果进行处理。

```
1   @Override
2    protected void onActivityResult(int requestCode,int resultCode,
3                                     Intentintent) {
4        super.onActivityResult(requestCode,resultCode,data);
5        if(requestCode = = 0 ){
6            if(resultCode = = Activity.RESULT_OK){
7                Bundle bundle = intent.getExtras();
8                 String info = bundle.getString("info","");
9                 Toast.makeText(MainActivity.this,info,
10                                  Toast.LENGTH_LONG).show();
11           }
12       }
13   }
```

onActivityResult( )方法的 3 个参数分别为请求码、结果码及携带返回数据的 intent 对象。在上述代码中，首先，判断请求码，区分是哪个请求返回的数据。如果请求码为 0，则表示是 SecondActivity 返回的数据。接着，判断结果码，不同的结果码处理数据的逻辑可能不同。如果结果码为"Activity. RESULT_OK"，则表示已接收到数据并返回，可以通过第 3 个参数 intent 对象获取返回数据。

【案例 1-12】 完成如下"自我介绍"界面，单击"编辑"按钮，跳转至"自我介绍编辑"页面，填写内容，单击"提交"按钮，回到"自我介绍"界面，将填写的内容返回并显示在下方，如图 1-41 所示。

图1-41 界面效果及页面跳转过程

1. 案例分析

案例中不仅需要从"自我介绍"页面跳转到"自我介绍编辑"页面,还需要将编辑页面中输入的内容返回"自我介绍"页面并显示。这是典型的向上一个Activity返回数据的情况。

2. 实现步骤

(1) 创建名为"SetResultDemo"的应用程序。

(2) 在activity_main.xml布局文件中完成自我介绍界面布局,具体代码可扫描二维码查看。

代码1:activity_main.xml

(3) 新建一个Activity,选择Empty Activity,Activity Name 设为"IntroduceActivity",Layout Name 设为"activity_introduce",并在生成的布局文件activity_introduce.xml中完成自我介绍编辑界面布局,具体代码可扫描二维码查看。

(4) 编写自我介绍界面中的功能逻辑代码。

在MainActivity.java中,实现单击"编辑"按钮跳转至自我介绍编辑界面,由于该界面需要将自我介绍内容返回至资料编辑界面,跳转时需要使用startActivityForResult()方法。同时,需要重写onActivityResult()方法,对返回的数据进行处理,具体代码如下。

代码2:activity_introduce.xml

```
1  public class MainActivity extends AppCompatActivity {
2
3      private Button bt_edit;
4      private TextView tv_intro;
5      public final static int REQUESTCODE = 1;//定义结果码常量
6      @Override
7      protected void onCreate(Bundle savedInstanceState) {
8          super.onCreate(savedInstanceState);
9          setContentView(R.layout.activity_main);
10         initView();
11     }
12     public void initView(){
13         bt_edit = findViewById(R.id.bt_edit);
```

```
14        tv_intro = findViewById(R.id.tv_intro);
15        bt_edit.setOnClickListener(new View.OnClickListener() {
16            @Override
17            public void onClick(View v) {
18                //创建 Intent 对象,此处使用显式 Intent
19                Intent intent = new Intent(MainActivity.this,
20                                          IntroduceActivity.class);
21                //启动 Activity,参数为 intent 及请求码
22                startActivityForResult(intent,REQUESTCODE);
23            }
24        });
25    }
26    //重写 onActivityResult 方法对返回结果进行处理
27    @Override
28    protected void onActivityResult(int requestCode,int resultCode,
29                                    @Nullable Intent data) {
30        super.onActivityResult(requestCode,resultCode,data);
31        //判断请求码
32        if(requestCode == REQUESTCODE){
33            //判断结果码
34            if(resultCode == RESULT_OK){
35                String introduction = data.getStringExtra("introduction");
36                tv_intro.setText(introduction);
37            }
38        }
39    }
40 }
```

(5) 编写自我介绍编辑界面的功能逻辑代码。

自我介绍编辑界面（IntroduceActivity.java）主要实现将自我介绍内容返回给前一个页面的功能，因此需要在单击"提交"按钮时将编辑框中的自我介绍取出，并通过 setResult() 方法返回，具体代码如下所示。

```
1  public class IntroduceActivity extends AppCompatActivity {
2
3      private Button bt_submit;
4      private EditText et_intro;
5      @Override
6      protected void onCreate(Bundle savedInstanceState) {
7          super.onCreate(savedInstanceState);
8          setContentView(R.layout.activity_introduce);
9          initView();
10     }
11     public void initView(){
12         bt_submit = findViewById(R.id.bt_submit);
```

```
13          et_intro = findViewById(R.id.et_intro);
14          bt_submit.setOnClickListener(new View.OnClickListener() {
15              @Override
16              public void onClick(View v) {
17                  String intrduction = et_intro.getText().toString();
18                  //创建 Intent 对象并放入需要返回的数据
19                  Intent intent = new Intent();
20                  intent.putExtra("introduction",intrduction);
21                  //调用 setResult()方法,参数为结果码及 intent 对象
22                  setResult(RESULT_OK,intent);
23                  finish();
24              }
25          });
26      }
27  }
```

注意：在调用 setResult( )方法后需要调用 finish( )方法将 IntroduceActivity 销毁；否则，IntroduceActivity 将依然处于活动状态，而数据信息已经传递给 MainActivity。

想一想：如果 IntroduceActivity.java 中，将上述代码中的第 19~22 行跳转部分进行如下改写，跟原来相比有何区别？

```
1  Intent intent = new Intent(MainActivity.this,IntroduceActivity.class);
2  intent.putExtra("introduction",intrduction);
3  startActivity(intent);
```

任务实施

1. 任务分析

在任务 2 及其后的巩固练习中，已经完成"我的信息"界面（MineActivity）布局及部分逻辑功能、"学习状态"界面（StatusActivity）布局，本任务需要在此基础上进一步实现如下功能。

（1）在"我的信息"界面（MineActivity）中，添加对"学习状态"的单击事件处理，跳转至"学习状态"界面，并将当前学习状态信息传至"学习状态"页面。

（2）在"学习状态"界面（StatusActivity）中，需要对每个状态进行单击事件监听，单击后将当前选择的学习状态返回给"我的信息"界面（MineActivity）。

2. 实现步骤

（1）在"我的信息"界面（MineActivity）中，添加功能逻辑代码，与前面相同的代码此处省略。

获取布局中学习状态相对布局（id 为"rl_status"）、状态图标控件（id 为"iv_status"）、状态文本控件（id 为"tv_status"）对象。学习状态相对布局对象添加单击事件监听，单击时跳转至"学习状态"界面，并将学习状态文本传至"学习状态"界面。由于此处需要接收"学习状态"界面返回的数据，需要调用 startActivityForResult( )方法实现界面跳转。

【代码1-9】 MineActivity.java

```java
public class MineActivity extends AppCompatActivity implements
                                        View.OnClickListener{
    //省略原变量定义语句
    //新增定义学习状态相对布局、状态图标及状态显示控件变量
    private RelativeLayout status;
    private ImageView iv_status;
    private TextView tv_status;

    @Override
    protected void onCreate(Bundle savedInstanceState) {
        super.onCreate(savedInstanceState);
        setContentView(R.layout.activity_mine);
        //初始化视图
        initView();
    }

    private void initView() {
        //省略原控件获取及设置事件监听代码
        //新增学习状态相关布局及控件获取代码
        status = findViewById(R.id.rl_status);
        iv_status = findViewById(R.id.iv_status);
        tv_status = findViewById(R.id.tv_status);
        //设置学习状态相对布局单击事件监听
        status.setOnClickListener(this);
    }
    @Override
    public void onClick(View v) {
        switch (v.getId()) {
            case R.id.user_name:
                usenameChangeDialog();   //调用昵称修改对话框方法
                break;
            case R.id.user_gender:
                genderChoiceDialog();    //调用性别选择对话框方法
                break;
            case R.id.user_birthday:
                birthdayDialog();        //调用出生日期选择对话框方法
                break;
            case R.id.rl_status:
                Intent intentToStatus = new Intent(MineActivity.this,
                                    StatusActivity.class);
                Bundle bundle = new Bundle();
                bundle.putString("now",tv_status.getText().toString());
                intentToStatus.putExtras(bundle);
                startActivityForResult(intentToStatus,2);
                break;
```

```
46            default:
47                break;
48        }
49    }
50    //省略对话框相关方法
51    //重写 onActivityResult()方法
52    @Override
53    protected void onActivityResult(int requestCode,
54                                    int resultCode,Intent data){
55        super.onActivityResult(requestCode,resultCode,data);
56        if (requestCode = =2 && resultCode = =RESULT_OK){
57            Bundle bundle = data.getExtras();
58            Status status = (Status) bundle.get("status");
59            int imageId = bundle.getInt("image");
60            String state = bundle.getString("state");
61            iv_status.setImageResource(imageId);
62            tv_status.setText(state);
63        }
64    }
65 }
```

(2) 在"学习状态"界面（StatusActivity）中，添加功能逻辑代码。

获取当前状态文本控件（id 为"tv_now_status"）、每个状态相对布局（id 为"rl_sick""rl_vacation""rl_study""rl_playing""rl_nothing"）对象，并为相对布局对象添加单击事件监听。在 onCreate( )方法中，获取 Intent 对象取出传过来的当前状态字符串，并显示在当前状态文本控件上。在单击每种学习状态时，需要将学习状态图标及文本放入 Intent 对象，通过 setResult( )方法返回。

【代码 1 – 10】 StatusActivity.java

```
1  public class StatusActivity extends AppCompatActivity implements
2                                    View.OnClickListener{
3      //获取控件
4      private TextView tv_now_status;
5      private RelativeLayout rl_sick;
6      private RelativeLayout rl_vacation;
7      private RelativeLayout rl_study;
8      private RelativeLayout rl_playing;
9      private RelativeLayout rl_nothing;
10
11     private Intent intent = new Intent();
12     private Bundle bundle = new Bundle();
13
14     @Override
15     protected void onCreate(Bundle savedInstanceState){
```

```java
16          super.onCreate(savedInstanceState);
17          setContentView(R.layout.activity_status);
18          //初始化视图
19          initView();
20          //获取Intent对象
21          Intent intent = getIntent();
22          //获取传来的Bundle对象
23          Bundle bundle = intent.getExtras();
24          //取出当前状态
25          String nowStatus = bundle.getString("now");
26          tv_now_status.setText(nowStatus);
27      }
28
29      private void initView() {
30          //找到对应的控件
31          tv_now_status = findViewById(R.id.tv_now_status);
32          rl_sick = findViewById(R.id.rl_sick);
33          rl_vacation = findViewById(R.id.rl_vacation);
34          rl_study = findViewById(R.id.rl_study);
35          rl_playing = findViewById(R.id.rl_playing);
36          rl_nothing = findViewById(R.id.rl_nothing);
37          //添加单击事件监听
38          rl_sick.setOnClickListener(this);
39          rl_vacation.setOnClickListener(this);
40          rl_study.setOnClickListener(this);
41          rl_playing.setOnClickListener(this);
42          rl_nothing.setOnClickListener(this);
43      }
44      @Override
45      public void onClick(View v) {
46          switch (v.getId()) {
47              case R.id.rl_sick:
48                  //将状态图标和文本放入Bundle对象
49                  bundle.putInt("image",R.drawable.sick);
50                  bundle.putString("state","生病中");
51                  break;
52              case R.id.rl_vacation:
53                  bundle.putInt("image",R.drawable.vacation);
54                  bundle.putString("state","放假中");
55                  break;
56              case R.id.rl_study:
57                  bundle.putInt("image",R.drawable.study);
58                  bundle.putString("state","学习中");
59                  break;
60              case R.id.rl_playing:
```

```
61            bundle.putInt("image",R.drawable.play);
62            bundle.putString("state","活动中");
63            break;
64        case R.id.rl_nothing:
65            bundle.putString("state","无");
66            break;
67     }
68     intent.putExtras(bundle);
69     setResult(RESULT_OK,intent);
70     finish();
71   }
72 }
```

任务反思

编写并运行程序，将在代码编写及程序调试过程中出现的异常信息、产生原因及解决方法记录在下方。

问题1：_____

产生原因：_____

_____

解决方法：_____

_____

_____

问题2：_____

产生原因：_____

_____

解决方法：_____

_____

任务总结及巩固

小白：Activity 就像"最熟悉的陌生人"，天天见，但也就是只知其一不知其二，背后的知识还挺多的。

师兄：Activity 是 Android 四大组件之一，也是我们可以直接看到的。它的状态变化以及回调生命周期方法的时机，会决定我们的代码写在哪里，因此非常重要。

小白：Intent 应该也很重要，有了它终于能让几个页面相互跳转了，感觉提升了一个等级。

> **师兄**：Intent 是组件间进行通信的重要工具，它就像一辆在城市间来回运行的大巴车，带着往返数据，在组件和组件间传递信息。我们现在只接触了 Intent 的一种用法，随着后续内容的学习，我们还将看到 Intent 在其他组件通信间发挥的作用。

一、基础巩固

1. 下列选项中，（　　）不是 Activity 的生命周期方法。

　A. onCreate( )　　　　　　　　B. onStart( )

　C. onDestroy( )　　　　　　　D. startActivity( )

2. 在启动一个 Activity 时，（　　）方法不会被调用。

　A. onCreate( )　　　　　　　　B. onStart( )

　C. onRestart( )　　　　　　　D. onResume( )

3. （多选）在单击回退键或调用 finish( )方法销毁一个 Activity 时，（　　）3 个方法会被调用。

　A. onPause( )

　B. onStart( )

　C. onStop( )

　D. onDestroy( )

4. 下列关于 Intent 的描述中，不正确的是（　　）。

　A. Intent 可以实现界面的跳转，也可以在不同组件间传递数据

　B. 使用显式 Intent 时不需要指明跳转的目的组件

　C. 使用隐式 Intent 时会设置 Action 用于进行目的组件的匹配

　D. Intent 是 Android 中组件通信的重要方式

5. 下列方法中，（　　）是在进行页面间数据传递时可能会用到的。

　A. startActivity( )

　B. startActivityForResult( )

　C. setResult( )

　D. onActivityResult( )

二、技术实践

利用本任务学习的内容完成如下登录功能。

登录页面（图 1-42）中用户类型分为个人用户和公司用户，初始登录密码为"111111"。

（1）在登录页面中，输入手机号码及登录密码，单击"登录"按钮，如果登录密码正确，则弹出图 1-43 所示的"登录成功"对话框。

图 1-42 登录界面　　　　　　　图 1-43 "登录成功"对话框

（2）单击登录页面中的"忘记密码"按钮，则跳转至忘记密码页面（图 1-44）；输入新密码和确认新密码，两个密码需一致；单击"获取验证码"按钮，弹出图 1-45 所示的对话框，显示手机号码及 6 位随机验证码。修改完成，单击"确定"按钮，返回登录页面。

图 1-44 "忘记密码"界面　　　　　　　图 1-45 "验证码"对话框

（3）如果修改过密码，则登录时需输入修改后的密码才可登录成功。

任务 4　用户信息的存储

任务描述

在任务 3 的基础上，进一步完成用户信息存储的功能。

在"我的信息"页面中，能够对昵称、性别、生日、学院及学习状态等信息进行修改，修改后需将用户信息进行保存；再次进入"我的信息"页面时，应显示已保存的用户信息。

任务学习目标

通过本任务需达到以下目标：
- 能够使用 SharedPreferences 对信息进行存储。
- 能够运用文件存储方式对数据进行存储。

技术储备

Android 对数据的存储主要有 3 种基本方式：简单存储（SharedPreferences）、文件存储以及 SQLite 数据库存储，这里我们主要讨论前两种存储方式。

一、简单存储（SharedPreferences）

（一）特点

SharedPreferences 保存基于 xml 文件存储的键值对（key – value）数据，通常用来保存少量且数据格式为字符串、整型等基本数据类型的数据，比如应用程序的各种配置信息（如是否打开音效、是否使用震动效果、小游戏的玩家积分等），解锁口令、密码等。

（二）存放位置

SharedPreferences 数据存储在 "/data/data/ < package name >/shared_prefs" 目录下，通过 Device File Explorer 面板可以查看，如图 1 – 46 所示。

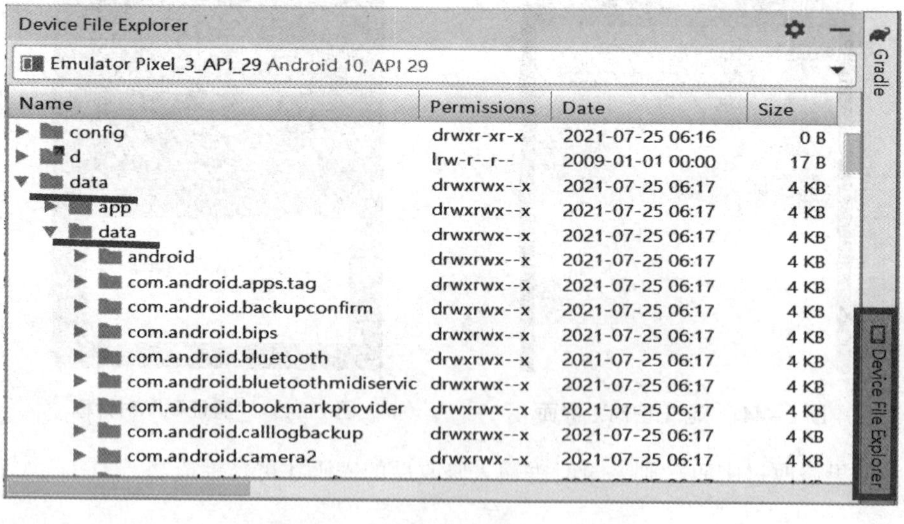

图 1 – 46 Device File Explorer 面板

（三）存取方式

1. 存数据

通过 SharedPreferences 存数据需要以下 3 个步骤。

（1）获取 SharedPreferences 对象。

SharedPreferences 本身是一个接口，程序无法直接创建 SharedPreferences 实例，只能通过 Context 提供的 getSharedPreferences( )方法来获取 SharedPreferences 实例,代码如下。

```
SharedPreferences spf = getSharedPreferences("data",MODE_PRIVATE);
```

其中,第 1 个参数为 xml 文件名,第 2 个参数为文件操作模式,其值为 Context 中提供的常量,分别如下。

① MODE_PRIVATE:只能被应用本身访问,在该模式下写入的内容会覆盖原文件的内容。

② MODE_APPEND:追加模式,该模式会检查文件是否存在,存在就往文件中追加内容,否则就创建新文件。

(2) 获取编辑器对象。

SharedPreferences 对象本身只能获取数据而不支持存储和修改,存储修改是通过 SharedPreferences. edit( )获取的内部接口 Editor 对象实现。

```
SharedPreferences.Editor editor = spf.edit();
```

(3) 将数据放入文件并提交。

通过 Editor 对象的 putXXX( )方法存放数据,putXXX( )方法的两个参数分别是 key(键) 和 value (值),并调用 commit( )方法提交数据。

```
1  editor.putString("name","张三");
2  editor.putInt("age",19);
3  editor.commit();
```

2. 取数据

取数据相对于存数据要简单一些,只需要如下两个步骤。

(1) 获取 SharedPreferences 对象,获取方法与存数据时相同。

(2) 通过 SharedPreferences 对象的 getXXX( )方法取数据。

```
SharedPreferences spf = getSharedPreferences("data",MODE_PRIVATE);
String name = spf.getString("name","");
int age = spf.getInt("age",18);
```

getXXX( )方法的第 1 个参数为要取数据的键,需要与存入时的键相同;第 2 个参数为缺省值,如果不存在该键,将返回缺省值。

注意:如果我们要删除 SharedPreferences 中的数据,可以调用 remove( )方法删掉某条数据,也可以调用 clear( )方法删除所有数据。

【案例 1-13】 在任务 1 中已经完成了学习通关登录界面的实现,现补充信息保存功能:在登录时,使用 SharedPreferences 保存手机号码及密码;下次登录时直接读取保存的手机号码显示在登录页面中,如图 1-47、图 1-48 所示。

图 1-47 保存后显示保存成功

图 1-48 再次打开显示号码

1. 案例分析

在本案例中，需要注意数据的存储及读取代码所在的位置。数据存储应在单击界面中的"登录"按钮，当手机号码和密码正确时进行，因此，应在"登录"按钮的单击事件监听方法 onClick()中实现数据存储；数据读取需要在创建登录界面时进行，如果已经保存过数据，则将保存的手机号显示在界面上，否则，手机号码处为空，因此，应在 onCreate()方法中实现数据读取。数据存储到 data.xml 文件中，data.xml 文件位置如图 1-49 所示。

图 1-49 data.xml 文件位置

2. 实现步骤

在任务 1 项目基础上进行功能实现。

（1）创建 SPSave 工具类。创建 saveUserInfo( ) 方法用于存储手机号码及密码；创建 getUserInfo( ) 方法用于读取数据。

```java
1  public class SPSave {
2      //存储手机号码及密码到 data.xml 文件中
3      public static void saveUserInfo(Context context,String phonenum,
4                                     String password){
5          SharedPreferences spf = context.getSharedPreferences("userData",
6                                  Context.MODE_PRIVATE);
7          SharedPreferences.Editor editor = spf.edit();
8          editor.putString("phonenum",phonenum);
9          editor.putString("password",password);
10         editor.commit();
11     }
12     //读取手机号码
13     public static Map<String,String> getUserInfo(Context context){
14         SharedPreferences spf = context.getSharedPreferences("userData",
15                                 Context.MODE_PRIVATE);
16         String phonenum = spf.getString("phonenum",null);
17         String password = spf.getString("password",null);
18         Map<String,String> useInfo = new HashMap<>();
19         useInfo.put("phonenum",phonenum);
20         useInfo.put("password",password);
21         return useInfo;
22     }
23 }
```

（2）在 MainActivity.java "登录"按钮的事件监听方法 onClick( ) 中，如果手机号码及密码正确，则调用 SPSave 工具类的 saveUserInfo( ) 存储数据，并用 Toast 显示"保存成功"提示信息。

（3）在 onCreate( ) 方法中，调用 SPSave 工具类的 getUserInfo( ) 存储数据读取 SharedPreferences 中的手机号码，并在手机号码输入框中显示。

```java
1  @Override
2  protected void onCreate(Bundle savedInstanceState) {
3      super.onCreate(savedInstanceState);
4      setContentView(R.layout.activity_main);
5      init();
6      //读取手机号码
7      Map<String,String> userInfo = SPSave.getUserInfo(this);
8      phonenum = userInfo.get("phonenum");
```

```
9        et_phonenum.setText(phonenum);
10       btn_login.setOnClickListener(new View.OnClickListener() {
11           @Override
12           public void onClick(View arg0) {
13               phonenum = et_phonenum.getText().toString();
14               password = et_password.getText().toString();
15               if(phonenum = = null || "".equals(phonenum)){
16                   Toast.makeText(MainActivity.this,"请输入手机号码",
17                           Toast.LENGTH_LONG).show();
18               }else if(password = = null || "".equals(password)){
19                   Toast.makeText(MainActivity.this,"请输入密码",
20                           Toast.LENGTH_LONG).show();
21               }else if(phonenum.equals("12345678910") &&
22                           password.equals("123456")){
23                   //存储用户名及密码
24                   SPSave.saveUserInfo(MainActivity.this,phonenum,password);
25                   Toast.makeText(MainActivity.this,"保存成功",
26                           Toast.LENGTH_LONG).show();
27                   //跳转至"我的信息"界面
28                   Intent intent =
29                       new Intent("com.example.learntopass.MineActivity");
30                   startActivity(intent);
31               }else{
32                 Toast.makeText(MainActivity.this,"手机号码或密码错误",
33                         Toast.LENGTH_LONG).show();
34               }
35           }
36       });
37   }
```

二、文件存储

文件存储是 Android 中最基本的一种数据存储方式，与 Java 文件存储类似，是通过 I/O 流的形式把数据存储到文件中。在 Android 中，根据文件存储位置的不同主要有两种类型：内部存储和外部存储，下面我们分别进行学习。

（一）内部存储

1. 特点

内部存储不同于内存，是图 1-50 中机身存储中的内部存储。内部存储中的文件默认只能被应用程序自己访问。当一个应用程序卸载之后，内部存储中的文件也将被删除。内部存储空间十分有限，它也是系统本身和系统应用程序主要的数据存储所在地，一旦内部存储空间耗尽，手机也就无法使用了，所以我们要尽量避免使用内部存储空间。

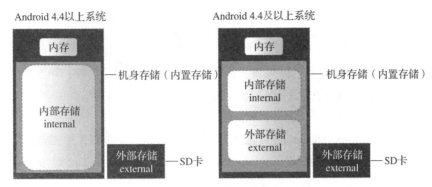

图 1-50 内部及外部存储示意图

2. 存放位置

内部存储的文件默认存储位置为"/data/data/<包名>/files"目录。我们可以通过下面的方法获取内部存储路径，见表 1-10。

表 1-10 获取内部存储路径的方法

方法	路径
Environment.getDataDirectory()	获取内部存储的根路径
getDataDir().getAbsolutePath()	获取应用在内部存储中的路径
getFilesDir().getAbsolutePath()	获取应用在内部存储中的 files 路径
getCacheDir().getAbsolutePath()	获取应用在内部存储中的 cache 路径

试一试：请尝试编写程序，记录以上 4 个方法返回的路径，观察获取的内部存储路径是否是默认路径？这些方法返回的路径有何区别？完成表 1-11。

表 1-11 方法返回的路径区别

方法	实际路径
Environment.getDataDirectory()	
getDataDir().getAbsolutePath()	
getFilesDir().getAbsolutePath()	
getCacheDir().getAbsolutePath()	

3. 存取方式

针对默认存储位置"/data/data/<包名>/files"目录，我们需要通过特定方法来获取 I/O 流对象。

（1）存储数据。

步骤 1：获取输出流对象。

输出流对象通过 Context 提供的 openFileOutputStream() 方法获取，方法具体定义形式如下。

```
FileOutputStream fos = openFileOutputStream(String name,int mode);
```

其中，第 1 个参数为文件名，如果该文件不存在，将自动创建文件；第 2 个参数为文件操作模式，其值可以为"MODE_PRIVATE"或"MODE_APPEND"。

步骤 2：通过输出流对象的 write() 方法存入数据。

我们创建 writeToInternalStorage() 方法实现文件的存储。方法有 3 个参数，第 1 个参数为上下文对象，通过该对象可以调用 openFileOutput() 方法获取 I/O 流对象，第 2 个参数为存储的文件名，第 3 个参数为要存储的数据。

```
1   public static void writeToInternalStorage(Context context,
2                          String fileName,String content){
3       FileOutputStream fos = null;
4       try{
5           //获取流对象
6           fos = context.openFileOutput(fileName + ".txt",MODE_PRIVATE);
7           //将数据通过流对象写出到文件
8           fos.write(content.getBytes());
9       }catch(Exception e){
10          e.printStackTrace();
11      }finally{
12          try{
13              //关闭流对象
14              if(fos! = null){
15                  fos.close();
16              }catch(IOException e){
17                  e.printStackTrace();
18              }
19      }}
```

(2) 读取数据。

步骤 1：获取输入流对象。

输入流对象通过 Context 提供的 openFileInputStream() 方法获取，方法具体定义形式如下，其参数为要读取的文件名。

```
FileInputStream fis = openFileInputStream(String name);
```

步骤 2：通过输入流对象的 read() 方法读取数据。

我们创建 readFromInternalStorage() 方法实现文件读取。方法有 3 个参数，第 1 个参数为上下文对象，通过该对象可以调用 openFileInput() 方法获取 I/O 流对象，第 2 个参数为读取的文件名，第 3 个参数为要读出的数据。

```
1   public static void readFromInternalStorage(Context context,
2                          String fileName,String content){
3       FileInputStream fis = null;
```

```
4       try {
5           fis = context.openFileInput(fileName + ".txt");
6           byte[] buffer = new byte[fis.available()];
7           fis.read(buffer);
8            content = new String(buffer);
9       } catch (Exception e) {
10          e.printStackTrace();
11      } finally {
12          try {
13              if (fis ! = null) {
14                  fis.close();
15              }
16          } catch(IOException e) {
17              e.printStackTrace();
18          }
19      }
20  }
```

(二) 外部存储

1. 特点

外部存储是指移动设备上的外部存储设备,比如我们常见的 SD 卡。在 Android 4.4 之后,很多设备中内嵌有存储卡,这些也是外部存储,如图 1-50 所示。外部存储中的文件可以被其他应用程序共享。

2. 存放位置

外部存储的文件通常位于 storage 目录下,但不同厂商生产的手机路径可能不同。我们可以通过以下方法获取外部存储路径。

(1) Environment. getExternalStorageDirectory():该方法返回外部存储根目录,其返回值为"File"。

(2) Environment. getExternalStorageDirectory(). getPath():该方法返回外部存储根路径,其返回值为"String"。

3. 存取方式

在进行外部存储文件存取时,通常需要通过上述方法获取外部存储路径,并通过 new 关键字创建 I/O 流对象,进行数据的读取和写入。

(1) 存储数据。

步骤 1:创建文件路径字符串。

下面代码所代表的文件路径为外部存储根目录"/self/data.txt"。

```
String filePath = Environment.getExternalStorageDirectory().getPath() + File.separator + "self" + File.separator + "data.txt ";
```

步骤 2:判断文件是否存在,如果不存在则创建文件夹及文件。

与内部存储不同,在外部存储中,当文件夹或文件不存在时,不会自动创建,需要在代码中进行判断并创建。我们创建 getFile( ) 方法实现文件的判断及创建，其参数为步骤1中创建的文件路径字符串。

```
1   public static File getFile(String filePath){
2       //通过文件路径创建文件对象
3       File file = new File(filePath);
4       //判断父文件夹是否存在,如果不存在则创建文件夹
5       if (! file.getParentFile().exists()) {
6           file.getParentFile().mkdirs();
7       }
8       //判断文件是否存在,如果不存在则创建文件
9       if (! file.exists()) {
10          try {
11              file.createNewFile();
12          } catch (Exception e) {
13              Log.e("创建文件失败",e.getMessage());
14          }
15      }
16      return file;
17  }
```

步骤3：创建输出流对象，进行数据的存储。

我们创建 writeToFile( ) 方法实现数据存储。方法的参数分别为文件路径字符串和要存储的数据。

```
1   public static void writeToFile(String filePath,String content){
2       //调用 getFile()方法创建文件并返回
3       File file = getFile(filePath);
4       FileWriter fw = null;
5       BufferedWriter bw = null;
6       try {
7           //创建缓冲流对象实现数据的写入
8           fw = new FileWriter(file,false);
9           bw = new BufferedWriter(fw);
10          bw.write(content);
11      } catch (Exception e) {
12          Log.e("写文件出错",e.getMessage());
13      } finally {
14          try {
15              //关闭流对象
16              if (bw ! = null)
17                  bw.close();
18              if (fw ! = null)
19                  fw.close();
20          } catch (IOException e) {
21              e.printStackTrace();
```

```
22          }
23        }
24    }
```

在上述代码中,调用了步骤 2 中创建的 getFile( )方法并返回文件对象;在文件存储时,使用缓冲流对象,能够更为方便地实现数据的存储。

(2) 读取数据。

步骤1:创建文件路径字符串。

本步骤与存储数据时创建文件路径字符串相同,此处不再赘述。

步骤2:创建输入流对象,进行数据的读取。

我们创建 readFile( )方法实现数据读取,方法的参数为文件路径字符串。

```
1   public static String readFile(String filePath) {
2       File file = new File(filePath);
3       String s = null;
4       StringBuilder stringBuilder = new StringBuilder();
5       FileReader fr = null;
6       BufferedReader reader = null;
7       int count = 0;
8       try {
9           //创建缓冲流对象
10          fr = new FileReader(file);
11          reader = new BufferedReader(fr);
12          //循环按行读取数据
13          while ((s = reader.readLine()) != null) {
14              stringBuilder.append(s);
15          }
16      } catch (Exception e) {
17          Log.e("读取文件出错",e.getMessage());
18      } finally {
19          try {
20              //关闭流对象
21              if (reader != null)
22                  reader.close();
23              if (fr != null)
24                  fr.close();
25          } catch (IOException e) {
26              e.printStackTrace();
27          }
28      }
29      return stringBuilder.toString();
30  }
```

在上述代码中,使用了缓冲流对象及 StringBuilder 对象,通过 BufferedReader 对象的 readLine( )方法能够便捷地实现按行读取数据,读取的数据追加到 StringBuilder 对象上,最后将 StringBuilder 对象转换为字符串返回,这是一个典型的文本文件读取方法。

前面我们分别创建不同的方法实现了内部存储及外部存储文件的写入和读取，在实际开发中，可以创建文件工具类 FileUtils，将创建的方法作为类的静态方法。这样就可以在需要进行文件读取的场合，通过 FileUtils 类调用所需方法，有效地实现了代码的复用。

4. 申请存储权限

为了保证应用程序的安全，Android 系统要求应用程序在访问系统关键信息时必须申请权限，否则程序运行时会因为没有访问权限而读写失败。由于外部存储可以被其他应用程序共享，也是不安全的，因此，在外部存储读写操作时，也需要进行存储权限申请，否则会出现文件创建失败、读写文件出错等问题。

根据 Android SDK 版本的不同，权限申请分为两种方式：静态权限申请和动态权限申请。

（1）静态权限申请。

静态权限申请的方式适用于 Android 6.0 以下的版本，这种方式是在 AndroidManifest.xml 文件中对所需权限进行声明。申请外部存储读写权限的代码如下所示。

```
1  <uses-permission android:name="android.permission.WRITE_EXTERNAL_STORAGE"/>
2  <uses-permission android:name="android.permission.READ_EXTERNAL_STORAGE"/>
```

此声明应置于 <manifest> </manifest> 节点之间，<application> 节点之外。

（2）动态权限申请。

Android 6.0 及以上版本的系统中，Android 将权限分为正常权限和危险权限。正常权限表示不会直接给用户隐私权带来风险的权限，比如请求网络的权限。危险权限表示涉及用户隐私的权限，申请该权限的应用可能涉及用户隐私数据，可能对用户存储的数据或其他应用的操作产生影响。危险权限一共分为9组，分别为存储（STORAGE）、照相机（CAMERA）、位置（LOCATION）、日历（CALENDAR）、联系人（CONTACTS）、传感器（SENSORS）、麦克风（MICROPHONE）、电话（PHONE）、短信（SMS），相关权限共24个。

申请正常权限时使用静态权限申请即可，而对于危险权限则需要用户授权后才可以使用，因此不仅需要在 AndroidManifest.xml 文件中添加权限，还需要在代码中进行动态权限申请。动态申请外部存储权限的步骤如下所示。

步骤1： 在 Activity 的 onCreate()方法中，添加动态权限申请。

```
1  if(ContextCompat.checkSelfPermission(this,
2                  Manifest.permission.WRITE_EXTERNAL_STORAGE)
3                  !=PackageManager.PERMISSION_GRANTED ||
4    ContextCompat.checkSelfPermission(this,
5                  Manifest.permission.READ_EXTERNAL_STORAGE)
6                  !=PackageManager.PERMISSION_GRANTED){
7       Log.e("权限问题","有权限吗");
8       ActivityCompat.requestPermissions(this,
9       new String[]{Manifest.permission.WRITE_EXTERNAL_STORAGE,
10      Manifest.permission.READ_EXTERNAL_STORAGE},1);
11  }else{
12      //文件存储代码
13  }
```

上述代码的 if 语句判断条件中，调用 checkSelfPermission( )方法判断当前应用程序是否已获得外部存储读、写权限，该方法包含 2 个参数，第 1 个参数为上下文对象，第 2 个参数为权限常量，"Manifest.permission.WRITE_EXTERNAL_STORAGE" 为外部存储写权限，"Manifest.permission.READ_EXTERNAL_STORAGE" 为外部存储读权限，也可以使用字符串"android.permission.WRITE_EXTERNAL_STORAGE" 和 "android.permission.READ_EXTERNAL_STORAGE"。常量 "PackageManager.PERMISSION_GRANTED" 代表有权限。

requestPermissions( )方法用于请求权限，该方法包含 3 个参数，第 1 个参数为上下文对象，第 2 个参数为需要申请的权限数组，第 3 个参数为请求码。运行程序时，界面上会弹出如图 1-51 所示的权限请求对话框，用户自主选择是否授予权限，"Allow"表示允许，"Deny"表示拒绝。

步骤 2： 添加回调方法 onRequestPermissionsResult( )。

当用户单击权限请求对话框中的按钮时，程序会执行动态权限申请的回调方法 onRequestPermissionsResult( )，在该方法中可以获取用户授予的权限结果，并针对不同结果完成不同操作。

图 1-51 权限请求对话框

```
1  @Override
2  public void onRequestPermissionsResult(int requestCode,String[] permissions,int[] grantResults){
3      super.onRequestPermissionsResult(requestCode,permissions,grantResults);
4      //判断请求码
5      if(requestCode = =1){
6          //循环判断权限授予结果
7          for (int i =0;i <permissions.length ;i + +){
8              if (grantResults[i] = =PackageManager.PERMISSION_GRANTED){
9                  Toast.makeText(this,"权限" +permissions[i] +"申请成功",
10                                 Toast.LENGTH_LONG).show();
11             }else{
12                  Toast.makeText(this,"权限" +permissions[i] +"申请失败",
13                                 Toast.LENGTH_LONG).show();
14             }
15         }
16     }
17 }
```

在上述代码中，onRequestPermissionsResult( )方法中包含 3 个参数，第 1 个参数为请求码，第 2 个参数为请求的权限数组，第 3 个参数为用户授予的权限结果；其中，第 2、3 个参数数组对应位置上的值分别代表了权限和这个权限的授予结果。在循环中，逐个判断权限授予的结果，并给出不同的响应。

> **注意**：从 Android 10.0 开始，在进行外部存储文件读写时，还需要在 Android Manifest.xml 文件的 application 节点中，加入以下属性设置。
>
> android:requestLegacyExternalStorage = " true"

【案例 1 – 14】 将登录页面中的手机号码和密码存储在内部存储文件 data.txt 中。

1. 案例分析

本案例的功能逻辑与【案例 1 – 13】基本相同，不同之处在于需将 SharedPreferences 的读取和写入代码改为通过文件存储实现。

2. 实现步骤

（1）创建工具类 FileUtils.java，用于实现数据的存储和读取功能，代码框架如下所示。

```java
public class FileUtils{
    //内部存储:存文件
    public static void writeToInternalStorage(Context context,
                            String fileName,String content){
        //具体代码与前面相同,故省略
    }
    //内部存储:取文件
    public static String readFromInternalStorage(Context context,
                            String fileName){
        //具体代码与前面相同,故省略
    }
}
```

（2）在 MainActivity.java "登录" 按钮的事件监听方法 onClick() 中，如果手机号码及密码正确，则调用 FileUtils 工具类的 writeToInternalStorage() 方法存储数据，并用 Toast 显示 "保存成功" 提示信息。

（3）在 onCreate() 方法中，调用 FileUtils 工具类的 readFromInternalStorage() 存储数据读取文件中的手机号码，并在手机号码输入框中显示。

```java
@Override
protected void onCreate(Bundle savedInstanceState){
    super.onCreate(savedInstanceState);
    setContentView(R.layout.activity_main);
    init();
    //读取手机号码
    String data = FileSave.readFromInternalStorage(this,"data");
    if(data! =null){
        et_phonenum.setText(data.split(",")[0]);
    }
    btn_login.setOnClickListener(new View.OnClickListener(){
        @Override
        public void onClick(View arg0){
            phonenum = et_phonenum.getText().toString();
```

```
15              password = et_password.getText().toString();
16              if(phonenum = = null || "".equals(phonenum)){
17                  Toast.makeText(MainActivity.this,"请输入手机号码",
18                          Toast.LENGTH_LONG).show();
19              }else if(password = = null || "".equals(password)){
20                  Toast.makeText(MainActivity.this,"请输入密码",
21                          Toast.LENGTH_LONG).show();
22              }else if(phonenum.equals("12345678910") &&
23                          password.equals("123456")){
24                  //存储用户名及密码
25                  String data = phonenum + "," + password;
26                  FileSave.writeToInternalStorage(MainActivity.this,
27                          "data",data);
28                  Toast.makeText(MainActivity.this,"保存成功",
29                      Toast.LENGTH_LONG).show();
30                  //跳转至"我的信息"界面
31                  Intent intent =
32                      new Intent("com.example.learntopass.MineActivity");
33                  startActivity(intent);
34              }else{
35                  Toast.makeText(MainActivity.this,"手机号码或密码错误",
36              Toast.LENGTH_LONG).show();
37              }
38          }
39      });
40  }
```

任务实施

1. 任务分析

本任务要求将"我的信息"页面中的用户信息进行保存,并在再次打开页面时显示保存信息。

"我的信息"页面的用户信息量较少,且可以表示成键值对的形式,因此可以采用 SharedPreferences 进行存储。

(1) 存储数据。

存入数据应在数据修改后进行保存。昵称、性别、生日都以弹出对话框的形式修改,存储数据代码应在对话框的事件监听中完成;学院信息是以 Spinner 下拉列表框的形式进行修改,存储数据代码应在 Spinner 的事件监听中;学习状态由"学习状态"界面返回,存储数据代码可以在 onActivityResult( )方法中。

(2) 读取数据。

读出并显示数据应在"我的信息"页面加载时完成,代码应在 onCreate( )方法中实现。

2. 实现步骤

在"我的信息"界面(MineActivity)中,添加功能逻辑代码,此处省略与前面相同的代码。

（1）创建本功能所需成员变量。

```
1  public class MineActivity extends AppCompatActivity implements
2                                    View.OnClickListener {
3      //省略原变量定义语句
4      private String userName;//用户的昵称
5      private String userGender;//用户的性别
6      private String userBirthday;//用户的出生日期
7      private String userCollege;//用户的学院
8      private int userStatusImageId;//用户状态图片ID
9      private String userStatus;//用户状态
10     private SharedPreferences sp;
11     private SharedPreferences.Editor editor;
12
13     //省略其他代码
14 }
```

（2）创建 initSharedPreferences() 方法用于进行 SharedPreferences 及 Editor 对象的初始化。

```
1  private void initSharedPreferences() {
2      //文件名为"userData.xml"
3      sp = getSharedPreferences("userData",MODE_PRIVATE);
4      //获取编辑器对象
5      editor = sp.edit();
6  }
```

（3）在昵称、性别、生日等信息的弹出对话框中，添加数据存储代码。

```
1  //昵称修改对话框
2  public void usenameChangeDialog(){
3      final View layout = View.inflate(this,R.layout.dialog_user_name,null);
4      AlertDialog.Builder builder = new AlertDialog.Builder(this);
5      builder.setTitle("修改昵称")
6              .setView(layout)
7              .setPositiveButton("确定",new DialogInterface.OnClickListener()
8              { @Override
9                public void onClick(DialogInterface dialog,int id) {
10                    EditText update_user_name =
11                            layout.findViewById(R.id.update_user_name);
12                    String newName =
13                            update_user_name.getText().toString().trim();
14                    //设置控件中的数据
15                    user_name_tx.setText(newName);
16                    //将获得的新昵称放入 SharedPreferences 中并提交
17                    editor.putString("userName",newName);
18                    editor.commit();
19                }
20             })
```

```java
21            .setNegativeButton("取消",new DialogInterface.OnClickListener()
22            {   @Override
23                public void onClick(DialogInterface dialog,int id){
24                }
25            });
26     AlertDialog dialog = builder.create();
27     dialog.show();
28  }
29  //性别选择单选对话框
30  public void genderChoiceDialog(){
31     final String[] genders = new String[]{"男","女"};
32     AlertDialog.Builder builder1 = new AlertDialog.Builder(this);
33     builder1.setTitle("性别选择")
34            .setSingleChoiceItems(genders,0,
35                new DialogInterface.OnClickListener() {
36                    @Override
37                    public void onClick(DialogInterface dialog,int which) {
38                        user_gender_tx.setText(genders[which]);
39                        //存入 SharedPreferences 中并提交
40                        editor.putString("userGender",genders[which]);
41                        editor.commit();
42                        dialog.dismiss();
43                    }
44                });
45     AlertDialog dialog1 = builder1.create();
46     dialog1.show();
47  }
48  //选择日期对话框
49  public void birthdayDialog() {
50       Calendar ca = Calendar.getInstance();
51       int  mYear = ca.get(Calendar.YEAR);
52       int  mMonth = ca.get(Calendar.MONTH);
53       int  mDay = ca.get(Calendar.DAY_OF_MONTH);
54       DatePickerDialog dialog1 = new DatePickerDialog(this,
55           new DatePickerDialog.OnDateSetListener() {
56               //日期选择器上的月份是从 0 开始的
57               @Override
58               public void onDateSet(DatePicker view,int year,
59                                int monthOfYear,int dayOfMonth)
60               {
61                   String birthday = year + "年" + (monthOfYear +1) +
62                                   "月" + dayOfMonth + "日";
63                   user_birthday_tx.setText(birthday);
64                   editor.putString("birthday",birthday);
65                   editor.commit();}
66           },mYear,mMonth,mDay);
67       //显示时间的对话框
```

```
68          dialog1.show();
69      }
70 }
```

(4) 在 initView( ) 方法中，添加 Spinner 控件的 OnItemSelectedListener 事件监听，并在 onItemSelected( ) 方法中添加数据存储代码。

```
1  private void initView() {
2      //省略原有代码
3      college_sp.setOnItemSelectedListener(new
4                          AdapterView.OnItemSelectedListener() {
5          @Override
6          public void onItemSelected(AdapterView<?> parent,View view,
7                                      int position,long id) {
8              //将选中项存入 SharedPreferences 中并提交
9              editor.putString("college",college[position]);
10             editor.commit();
11         }
12         @Override
13         public void onNothingSelected(AdapterView<?> parent) {
14         }
15     });
16 }
```

(5) 在 onActivityResult( ) 方法中，添加学习状态图标及文本存储代码。

```
1  @Override
2  protected void onActivityResult(int requestCode,int resultCode,
3                                   Intent data) {
4      super.onActivityResult(requestCode,resultCode,data);
5      if (requestCode = = 2 && resultCode = = RESULT_OK) {
6          //省略原有代码
7          editor.putInt("imageId",imageId);
8          editor.putString("status",state);
9          editor.commit();
10     }
11 }
```

(6) 在 initSharedPreferences( ) 方法中，添加数据读取的代码，并在 onCreate( ) 方法中调用该方法。

```
1  private void initSharedPreferences() {
2      //文件名为"userData.xml"
3      sp = getSharedPreferences("userData",MODE_PRIVATE);
4      //获取编辑器对象
5      editor = sp.edit();
6      //通过对应的键存取的值,第2个参数为默认值
7      userName = sp.getString("userName","小白");
```

```
8      userGender = sp.getString("userGender","男");
9      userBirthday = sp.getString("birthday","未设置");
10     userCollege = sp.getString("college","智慧学院");
11     userStatusImageId = sp.getInt("imageId",R.drawable.study);
12     userStatus = sp.getString("status","学习中");
13  }
```

（7）在 setSpfValues()方法中，将读出的用户信息设置到相应控件上，并在 onCreate()方法中调用该方法。

```
1   private void setSpfValues(){
2       user_name_tx.setText(userName);//昵称初始值
3       user_gender_tx.setText(userGender);//性别初始值
4       user_birthday_tx.setText(userBirthday);//出生日期初始值
5       //学院初始值
6       for(int i = 0;i < college.length;i + +){
7           if(userCollege.equals(college[i])){
8               college_sp.setSelection(i);
9               break;
10          }
11      }
12      //状态保存值
13      iv_status.setImageResource(userStatusImageId);
14      tv_status.setText(userStatus);
15  }
```

任务反思

编写并运行程序，将在代码编写及程序调试过程中出现的异常信息、产生原因及解决方法记录在下方。

问题1：

产生原因：

解决方法：

问题2：

产生原因：

解决方法：

任务总结及巩固

小白：师兄，数据存储是每一次学习一门编程语言都绕不过的一部分内容呀。

师兄：这是当然，数据需要持久化，那就需要进行存储。你能理清数据存储这部分的思路吗？

小白：数据存储虽然有多种形式，但是在学习的时候对于每种方法不外乎要掌握以下几点。

- 存储特点是什么？这决定了此种存储方式适用的场合。
- 存成什么形式？存在哪里？
- 数据怎么存？数据怎么取？

师兄：非常好，学习中一定要善于总结，并做好知识的迁移，这次学习的存储方式中的文件存储其实跟 Java 中学习的一样，只不过在内部存储时获取流对象的方式不一样，外部存储时，要注意存储路径，同时，需要获得外部存储的访问权限。

小白：说到权限，前几天我看到一条新闻说有不少 App 因为强制、频繁、过度索取权限被通报，要求限期整改呢。

师兄：没错，需要索取的权限都是涉及用户信息安全和隐私权的，过度索取极有可能造成用户信息泄露，还有些应用会违规收集用户个人信息，这都是侵害用户权益的行为。国家也出台了《网络安全法》《App 违法违规收集使用个人信息行为认定方法》等法律法规来规范开发者的行为，保护用户的合法权益。我们作为开发人员也要了解相关法律法规，守好我们的底线。

一、基础巩固

1. 下列关于 SharedPreferences 存取文件的描述中，错误的是（　　）。

A. SharedPreferences 中存储的是键值对的形式

B. SharedPreferences 存储数据的保存形式是 xml 文件

C. SharedPreferences 存储的地址是 "/storage/shared_prefs"

D. SharedPreferences 可以用来存储用户的偏好设置

2. 下列方法中，（　　）方法可以用来获取 SharedPreferences 的编辑器对象。

A. edit( )　　　　　　　　　　B. editor( )

C. getEdit( )　　　　　　　　　D. putEdit( )

3. 下列方法中，（　　）方法可以用来获取内部存储输入流对象。

A. new FileInputStream( )　　　B. new FileOutputStream( )

C. openFileInput( )　　　　　　D. openFileOutput( )

4. 下列 Android 版本中，（　　）开始针对危险权限需要动态权限申请。

A. Android 4.4　　　　　　　　B. Android 6.0

C. Android 8.0　　　　　　　　D. Android 10

5. 下列代码中，（　　）语句可以创建 SharedPreferences 对象。

A. SharedPreferences sp = new SharedPreferences( )；

B. SharedPreferences sp = context. SharedPreferences( )；

C. SharedPreferences sp = getSharedPreferences( )；

D. SharedPreferences sp = SharedPreferences. getManager( )；

二、技术实践

参考如下界面效果（图 1 – 52、图 1 – 53、图 1 – 54、图 1 – 55），利用本任务中学习的内容完成一个记事本，具体功能可参考如下要求。

图 1 – 52　记事本首页

图 1 – 53　注册界面

首次打开记事本需要进行注册，注册完成后，跳转至登录页面，勾选"记住密码"选项则将密码进行存储；再次打开记事本，"注册"按钮改为"登录"按钮，单击直接进入登录界面，并直接填入用户名和密码；若用户名和密码正确，则进入记事页面；输入标题及内容，选择存储位置，单击"保存"按钮，以文件形式进行存储。

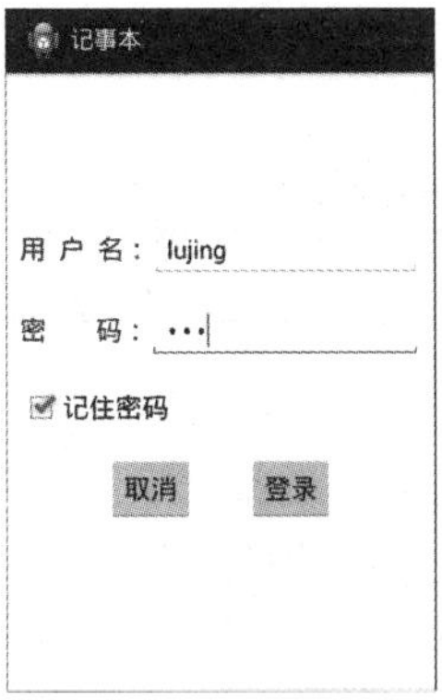

图 1 – 54　登录界面

图 1 – 55　记事页面

学习目标达成度评价

了解对本模块的自我学习情况,完成表 1-12。

表 1-12 学习目标达成度评价表

序号	学习目标	学生自评
1	能够使用基本控件 Button、TextView、EditText 等进行页面功能实现	□能够熟练使用基本控件 □通过查看以往代码及课本,能够基本实现基本控件的应用 □不会控件的应用
2	能够使用常用布局 LinearLayout、RelativeLayout 进行页面布局	□能够熟练使用并实现布局效果 □通过查看以往代码及课本,能够基本实现布局 □不会布局的分析及实现
3	能够使用 selector、shape、style 等进行界面效果实现	□能够熟练使用并实现布局效果 □通过查看以往代码及课本,能够基本实现界面效果 □不会使用
4	能够理解 Activity 生命周期并熟知生命周期状态变化及需回调的方法	□对本部分内容理解并熟知 □能够理解 □不能理解
5	能够应用 SharedPreferences 对数据进行存储	□能够熟练运用 □通过查看以往代码及课本,能够基本实现 □不知道如何存储
6	能够应用 I/O 流进行内部及外部文件存储	□能够熟练运用 □通过查看以往代码及课本,能够基本实现 □不知道如何存储
	评价得分	

学生自评得分(20%)	学习成果得分(60%)	学习过程得分(20%)	模块综合得分

(1) 学生自评得分。

每个学习目标有 3 个选项,选择第 1 个选项得 100 分,选择第 2 个选项得 70 分,选择第 3 个选项得 50 分,学生自评得分为各项学习目标得分的平均分。

(2) 学习成果得分。

教师根据学生阶段性测试及模块学习成果的完成情况酌情赋分,满分为 100 分。

(3) 学习过程得分。

教师根据学生的其他学习过程表现,如到课情况、参与课程讨论等情况酌情赋分,满分为 100 分。

功能模块 2

限时答题功能的实现

说在前面

小白：师兄，通过上一个功能模块感觉 Android 好像更重于应用呀，知道某个组件的功能和用法，在具体功能中进行应用就可以了。

师兄：我们现在学习的 Android 开发以 Java 作为基础，封装度是比较高的，如果深究底层实现，那学问更多了，你可以试着去找找比如 TextView 类的源码看看。我们现在是初学阶段，重点是要学会如何应用各种组件，并且能够自己分析界面效果可以通过哪些组件来实现、如何实现，来实现更为丰富的功能。

小白：下面我们要完成什么模块呢？

师兄：我们的 App 叫"学习通关"，要"通关"自然要答题啦。我们下面要攻克的就是限时答题功能，实现这个模块我们要循序渐进地完成以下 4 个任务。

任务 1：学习积分页面的设计与实现。

任务 2：答题功能页面的设计与实现。

任务 3：使用数据库存储题目信息。

任务 4：倒计时功能的实现。

功能需求描述

（1）进入如图 2-1 所示的学习积分列表界面，列表中显示学习项目及当前总积分。

（2）单击每日答题中的"去看看"按钮，进入如图 2-2、图 2-3 所示的限时答题界面。题目类型包括单选题和多选题，选择后单击"提交"按钮，会根据答案显示是否正确；单击"下一题"按钮，可继续答题；答题完成后显示总得分。

（3）答题限时 2 分钟，在答题页面左上角设置倒计时，显示答题剩余时间。

图 2-1　积分列表界面　　　图 2-2　单选答题界面　　　图 2-3　多选答题界面

学习目标

1. 知识目标

（1）列表控件（ListView）及适配器（SimpleAdapter、BaseAdapter）的使用方法。

（2）单选按钮（RadioButton）、复选按钮（CheckBox）及碎片（Fragment）的使用方法。

（3）SQLite 数据库的使用方法。

（4）内容提供者（ContentProvider）的使用方法。

2. 技能目标

（1）综合运用 ListView、SimpleAdapter、BaseAdapter 等设计并实现包含列表的界面。

（2）综合运用单、复选控件、碎片及其他控件实现相对复杂的界面效果及功能。

（3）能够通过 SQLite 数据库实现数据的存储。

（4）能够使用 ContentProvider 实现数据共享。

任务1　学习积分页面的设计与实现

任务描述

根据前面的参考界面，完成图 2-1 界面中的学习积分页面，单击学习项目"每日答题"后面的"去看看"按钮，能够跳转到答题页面。

任务学习目标

通过本任务需达到以下目标：

➢ 能够灵活使用 ListView 展示数据。

➢ 能够恰当运用 SimpleAdapter 以及 BaseAdapter 自定义适配器。

技术储备

一、ListView 的应用

（一）作用

ListView 是列表视图，其显示效果如图 2-4 所示，它允许在页面上分行展示相似的数据，如新闻列表、商品列表、书籍列表等，方便用户逐行浏览与操作，是我们在应用中常见的一种控件。

（二）基本用法

ListView 与前面学习的 Spinner 都是 AdapterView 的间接子类。在使用时，需要在布局文件中加入 ListView 控件，具体代码如下所示。

```
1  <LinearLayout xmlns:android = "http://schemas.android.com/apk/res/android"
2      android:layout_width = "match_parent"
3      android:layout_height = "match_parent"
4      android:orientation = "vertical"
5      android:padding = "20dp" >
6      <ListView
7          android:id = "@+id/listview"
8          android:layout_width = "match_parent"
9          android:layout_height = "wrap_content"/>
10 </LinearLayout>
```

图 2-4 ListView 显示效果

要实现在 ListView 中显示数据，还需要借助适配器完成数据与 ListView 视图的关联。如果 ListView 中显示的列表项内容是纯文本，则可以使用前面介绍的 ArrayAdapter 进行数据适配，但当列表项中包含图片等更为复杂的数据及布局时，则需要使用更为强大的适配器。

（三）简单适配器（SimpleAdapter）

SimpleAdatpter 的构造方法如下所示。

```
    public SimpleAdapter(Context context,List <? extends Map <String,?>>data,
int resource,String[]from,int[] to)
```

SimpleAdatpter 的构造方法有 5 个参数，具体含义如下所示。

- Context context：上下文对象。

- List <? extends Map <String,? >>data：数据集合，data 中的元素需要是 Map 对象，一个 Map 对象对应于 ListView 中的一个列表项数据。

- int resource：列表项布局的资源 ID。

- String [ ] from：Map 集合里面的键。

- int [ ] to：列表项布局相应的控件 ID。

（四）ListView 的事件处理

当 ListView 的列表项被单击时，会触发 OnItemClickListener 监听器。我们可以通过如下方法为 ListView 设置事件监听。

```
1  list.setOnItemClickListener(new AdapterView.OnItemClickListener() {
2      @Override
3      public void onItemClick(AdapterView<?> adapterView,View view,int position,long 1) {
4          //单击列表项的具体操作代码
5      }
6  });
```

单击列表项时的具体功能逻辑代码写在 onItem-Click()方法中，该方法的第3个参数为单击列表项的位置，也就是在数据集合中的索引。

【案例 2-1】 实现如图 2-5 所示的行星列表。

1. 案例分析

本案例界面效果是典型的 ListView 列表显示效果。每个列表项包含图片和文字，需要单独编写列表项布局。可以使用 SimpleAdapter 进行列表项数据加载，使用时需要将每一个列表项数据放入 Map 对象中，并将 Map 对象添加到 List 集合中。

2. 实现步骤

（1）创建名为"SimpleListViewDemo"的程序。

（2）导入案例所需图片。

图 2-5 行星列表

在 res 目录中，创建"drawable-xxhdpi"文件夹，将本案例所需图片放入该文件夹下，具体创建方法如下。

选中 res 目录，右击，依次选择 New→Android Resource Directory，在弹出的对话框中按图 2-6 所示进行设置，单击">>"按钮，在右侧选择"XX-High Density"，单击"OK"按钮。

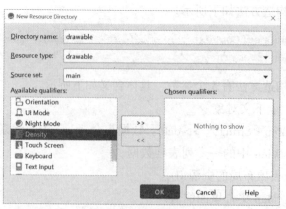

图 2-6 New Resource Directory 对话框

（3）编写列表项布局。

本列表的列表项中包含图片和文字，需要自定义列表项布局。在 res/layout 目录下创建名为"item_list.xml"的布局文件。具体代码可扫描二维码查看。

(4) 在主界面布局文件"activity_main.xml"中，加入 ListView 控件，具体代码与前面相同。

(5) 在 MainActivity.java 中编写功能逻辑代码。

代码：item_list.xml

```java
1  public class MainActivity extends AppCompatActivity {
2      private ListView list;
3      private List<Map<String,Object>> data;
4      private int[] iconArray =
5              {R.mipmap.shuixing,R.mipmap.jinxing,R.mipmap.diqiu,
6               R.mipmap.huoxing,R.mipmap.muxing,R.mipmap.tuxing};
7      private String[] starArray ={"水星","金星","地球","火星",
8                                    "木星","土星"};
9      @Override
10     protected void onCreate(Bundle savedInstanceState) {
11         super.onCreate(savedInstanceState);
12         setContentView(R.layout.activity_main);
13         list = findViewById(R.id.listview);
14         data = new ArrayList<>();
15         //将列表项数据组织为 List 对象
16         for (int i=0;i<iconArray.length;i++) {
17             Map<String,Object>item = new HashMap<String,Object>();
18             item.put("icon",iconArray[i]);
19             item.put("name",starArray[i]);
20             data.add(item);
21         }
22         //创建适配器对象
23         SimpleAdapter starAdapter = new SimpleAdapter(this,data,
24             R.layout.item_list,new String[]{"icon","name"},
25             new int[]{R.id.iv_icon,R.id.tv_name});
26         //设置适配器
27         list.setAdapter(starAdapter);
28     }
29     //设置事件监听
30     list.setOnItemClickListener(new AdapterView.OnItemClickListener() {
31         @Override
32         public void onItemClick(AdapterView<?>adapterView,View view,
33                                 int position,long l) {
34             Toast.makeText(MainActivity.this,
35                 "您单击的是"+starArray[position],Toast.LENGTH_LONG).show();
36         }
37     });
38 }
```

下面对 MainActivity 中的主要代码进行分析。

第 4~8 行为数据准备，每个列表项中包含图标及文字两种数据，分别将其组织为数组 iconArray 和 starArray。

第 3 行定义数据引用变量，并在第 14 行创建对象。需按照 SimpleAdapter 参数要求，定义为 "List < Map < String，Object > >" 类型，即定义为一个 List 集合，集合中的每一个元素为一个 Map 对象。

第 16 ~ 21 行将数据放入集合中，每一个列表项中的两种数据组织为一个 Map 对象，图标数据的键为 "icon"，值则从 iconArray 数组中取出，文字数据的键为 "name"，值则从 starArray 数组的相应位置取出。

第 23 ~ 26 行定义 SimpleAdater 对象，注意最后一个参数为控件 ID 组成的数组，其 ID 来自于列表项布局 item_list. xml，并且数组中控件 ID 的顺序应与第 4 个参数中键所代表的值所对应的控件一致。

> **长知识：**
>
> **Android 中的单位**
>
> 在 Android 开发中，我们会遇到诸如 dp、sp、dpi、px 等多种单位。其中，px 代表像素，也就是屏幕上的一个点；dpi 是每英寸的像素数，也叫屏幕密度，值越大，屏幕就越清晰；dp 是与密度无关的像素，在 160 dpi 的屏幕上，1 dpi 等于 1 px，一般在界面布局中使用；sp 是与缩放无关的像素，一般用来设定布局中的文字大小。
>
> 在图片资源目录中，会有 drawable_hdpi、drawable_xhdpi 等不同图片文件夹，Android 系统会根据设备屏幕密度自动匹配不同文件夹中的图片资源。一般 hdpi 匹配的密度范围为 160 ~ 240 dpi、xhdpi 为 240 ~ 320 dpi、xxhdpi 为 320 ~ 480 dpi。

二、基本适配器（BaseAdapter）

前面介绍的两种适配器 ArrayApater 及 SimpleAdapter 都有一个共同的父类 BaseAdapter。BaseAdapter 是一个抽象类，当列表项中存在多个控件，形成更为复杂的列表时，我们也可以通过继承 BaseAdapter 实现自定义适配器。

BaseAdapter 中有 4 个抽象方法，在实现自定义适配器时，需要实现这 4 个方法，并通过这些方法对 ListView 控件进行数据适配。

查找 API 文档，学习 BaseAdapter 中的 4 个抽象方法的功能，完成表 2 – 1。

表 2 – 1 BaseAdapter 的方法

方法名称	功能描述
public int getCount( )	
public Object getItem（int position）	
public long getItemId（int position）	
public View getView（int position，View ConvertView，ViewGroup parent）	

【案例 2-2】 创建自定义适配器实现【案例 2-1】中的界面效果。

1. 案例分析

在案例中,行星图片及行星名字共同构成了一个行星对象,可以将这些属性进行抽象,形成行星类 Planet,这样更好地体现了面向对象的思想。

2. 实现步骤

(1) 创建名为 "BaseListViewDemo" 的程序。

(2) 编写主界面布局及列表项布局。具体代码与【案例 2-1】中相同,此处不再赘述。

(3) 创建实体类 Planet。

```
1  public class Planet {
2      //成员变量
3      private int image;      //行星图片
4      private String name;//行星名字
5      //省略构造方法及 getter、setter 方法
6  }
```

(4) 创建 PlanetAdapter 继承 BaseAdapter。

```
1   public class PlanetAdapter extends BaseAdapter {
2       private int resource;      //列表项布局
3       private List<Planet> data;    //数据集合
4       private Context context;
5       //构造方法
6       public PlanetAdapter(Context context,int resource,List<Planet> data) {
7           this.resource = resource;
8           this.data = data;
9           this.context = context;
10      }
11      //获取列表项的个数
12      @Override
13      public int getCount() {
14          return data.size();
15      }
16      //获取 position 位置上的列表项对象
17      @Override
18      public Object getItem(int position) {
19          return data.get(position);
20      }
21      //获取 position 位置上的列表项的 id
22      @Override
23      public long getItemId(int position) {
24          return position;
25      }
26      //获取 position 位置上的列表项视图
27      @Override
```

```java
28      public View getView(int position,View convertView,ViewGroup parent){
29          //获取position位置上的数据对象
30          Planet planet = data.get(position);
31          //加载item_list.xml布局文件
32          View view = View.inflate(context,resource,null);
33          ImageView iv_icon = view.findViewById(R.id.iv_icon);
34          TextView tv_name = view.findViewById(R.id.tv_name);
35          //将数据设置到控件上
36          iv_icon.setImageResource(planet.getImage());
37          tv_name.setText(planet.getName());
38          return view;
39      }
40  }
```

（5）在 MainActivity 中编写功能逻辑代码。

```java
1   public class MainActivity extends AppCompatActivity{
2       private List<Planet> planets;
3       @Override
4       protected void onCreate(Bundle savedInstanceState){
5           super.onCreate(savedInstanceState);
6           setContentView(R.layout.activity_main);
7           initData();//初始化数据
8           ListView listview = findViewById(R.id.listview);
9           //创建MyAdapter对象
10          MyAdapter adapter = new MyAdapter(this,planets,R.layout.item_list);
11          //设置Adapter
12          listview.setAdapter(adapter);
13      }
14      //初始化数据
15      private void initData(){
16          planets = new ArrayList<Planet>();
17          Planet planet1 = new Planet(R.drawable.diqiu,"地球");
18          Planet planet2 = new Planet(R.drawable.tuxing,"土星");
19          Planet planet3 = new Planet(R.drawable.shuixing,"水星");
20          planets.add(planet1);
21          planets.add(planet2);
22          planets.add(planet3);
23      }
24  }
```

下面对 PlanetAdapter 代码进行分析。

在第 6~11 行的构造方法中，定义 3 个参数分别代表上下文对象 Context、列表项数据集合及列表项布局 id。

第 28~40 行实现了 getView() 方法。在实现 getView() 方法时主要完成两项工作：找到列表项布局中的控件及把数据放到控件上。第 37~38 行将数据从 Planet 对象中取出设置到控件上。

任务实施

1. 任务分析

本任务中的学习积分界面可以分为如图2-7所示的4个部分，整体可以采用垂直线性布局，ListView控件位于最下方。列表项需要单独实现一个列表项布局，并自定义适配器实现ListView控件与数据的适配。

2. 实现步骤

在功能模块1中实现的LearnToPass项目中，继续实现下列步骤。

(1) 在包com.project.learntopass中，创建新的包，具体包名为"activity""adapter""entity"。将原Activity放入包activity中。

图2-7 学习积分布局分析

(2) 在包activity中，创建新的EmptyActivity，设置Activity Name为"PointsActivity"，Layout Name为"activity_points"。

(3) 在activity_points.xml文件中，编写界面布局代码，具体代码可扫描二维码查看。

【代码2-1】 activity_points.xml

代码：activity_points.xml

```
1   <?xml version="1.0" encoding="utf-8"?>
2   <LinearLayout xmlns:android="http://schemas.android.com/apk/res/android"
3       xmlns:app="http://schemas.android.com/apk/res-auto"
4       xmlns:tools="http://schemas.android.com/tools"
5       android:layout_width="match_parent"
6       android:layout_height="match_parent"
7       android:orientation="vertical"
8       tools:context=".activity.PointsActivity">
9       <!--省略顶部部分布局-->
10      <!--省略总积分部分布局-->
11      <!--省略今日积分部分布局-->
12      <!--学习项目列表-->
13      <ListView
14          android:id="@+id/points_lv"
15          android:layout_width="match_parent"
16          android:layout_height="match_parent"
17          android:layout_marginTop="5dp" />
18  </LinearLayout>
```

(4) 在res/layout目录中，创建item_points.xml文件，编写列表项布局，具体代码可扫描二维码查看。

(5) 在包entity中，创建名为"PointsItem"的实体类。该实体类对

代码：item_points.xml

象代表一个学习项目，其属性包括项目标题及计分方法。

【代码 2-2】 PointsItem. java

```
1   public class PointsItem{
2       private String title;//项目标题
3       private String introduce;//积分方法介绍
4       public PointsItem(String title,String introduce){
5           this.title=title;
6           this.introduce=introduce;
7       }
8       //省略 getter、setter 方法
9   }
```

（6）在包 adapter 中，创建名为"PointsAdapter"的自定义适配器类。

【代码 2-3】 PointsAdapter. java

```
1   public class PointsAdapter extends BaseAdapter{
2       private List<PointsItem>list;//数据
3       private Context context;//上下文对象
4       //构造方法接收两个参数:上下文对象及数据集合
5       public PointsAdapter(Context context,List<PointsItem>list){
6           this.list=list;
7           this.context=context;
8       }
9       @Override
10      public int getCount(){
11          return list.size();
12      }
13      @Override
14      public Object getItem(int position){
15          return list.get(position);
16      }
17      @Override
18      public long getItemId(int position){
19          return position;
20      }
21      @Override
22      public View getView(int position,View convertView,ViewGroup parent){
23          final int p=position;
24          //获取当前数据对象
25          PointsItem item=list.get(position);
26          //加载列表项布局文件并获取控件对象
27          View view=View.inflate(R.layout.points_item,parent,false);
28          TextView points_title=view.findViewById(R.id.points_title);
29          TextView points_introduce=view.findViewById(R.id.points_introduce);
30          Button points_button=view.findViewById(R.id.points_button);
31          //设置数据
```

```
32      points_title.setText(item.getTitle());
33      points_introduce.setText(item.getIntroduce());
34      return view;
35    }
36  }
```

(7) 在 PointsActivity.java 中，编写功能逻辑代码。

【代码 2-4】 PointsActivity.java

```
1  public class PointsActivity extends AppCompatActivity{
2    List < PointsItem > list = new ArrayList < >();
3    @Override
4    protected void onCreate(Bundle savedInstanceState) {
5        super.onCreate(savedInstanceState);
6        setContentView(R.layout.activity_points);
7        //准备数据
8        initData();
9        //找到对应的控件
10       ListView points_lv = findViewById(R.id.points_lv);
11       PointsAdapter adapter = new PointsAdapter(this,list);
12       points_lv.setAdapter(adapter);
13       //添加列表项单击事件
14       points_lv.setOnItemClickListener(new AdapterView.OnItemClickListener(){
15           @Override
16           public void onItemClick(AdapterView<?> parent,View view,
17                                  int position,long id) {
18               Toast.makeText(PointsActivity.this,"列表项被单击了",
19                              Toast.LENGTH_SHORT).show();
20           }
21       });
22    }
23    private void initData() {
24        list.add(new PointsItem("登录","1分/每日首次登录"));
25        list.add(new PointsItem("时政学习","1分/每日有效阅读、播报一篇"));
26        list.add(new PointsItem("视听学习","1分/每日有效收听、观看一个"));
27        list.add(new PointsItem("每日答题","1分/每组答题每答对1道积1分"));
28        list.add(new PointsItem("挑战答题","每日仅前两局得分;每日上限5分"));
29    }
30  }
```

(8) 在包 activity 中，创建 QuestionActivity，其布局文件名为"activity_question"。

(9) 在 PointsAdapter.java 中为"每日答题"学习项目中的"去看看"按钮添加单击事件，跳转至 QuestionActivity 页面。

【代码 2-5】 PointsAdapter.java

```
1  public class PointsAdapter extends BaseAdapter {
2      //省略原代码
```

```
3      @Override
4      public View getView(int position,View convertView,ViewGroup parent){
5          final int p = position;
6          //省略原代码
7          points_button.setOnClickListener(new View.OnClickListener(){
8              @Override
9              public void onClick(View v){
10                 //判断当前位置position,如果是3,则代表"每日答题"项目
11                 switch(p){
12                     case 3:
13                         Intent intent = new Intent(context,
14                                         QuestionActivity.class);
15                         context.startActivity(intent);
16                         break;
17                     default:break;
18                 }
19             }
20         });
21         //设置数据
22         points_title.setText(item.getTitle());
23         points_introduce.setText(item.getIntroduce());
24         return view;
25     }
26  }
```

> **注意**：列表项中的按钮单击事件会与列表项单击事件发生冲突，造成列表项单击事件失效。可在列表项的按钮控件中，添加属性 android：focusable = "false"，避免按钮抢占列表项的焦点。

任务反思

编写并运行程序，将在代码编写及程序调试过程中出现的异常信息、产生原因及解决方法记录在下方。

问题1：_____

产生原因：_____

解决方法：_____

问题2：_____

产生原因：_____

_____

解决方法：_____

_____

任务总结及巩固

一、基础巩固

1. 下列方法中，(   ) 可以用来为 ListView 设置列表项单击事件监听。

  A. setOnClickListener( )　　　　　　B. setOnItemClickListener( )

  C. setOnItemSelectedListener( )　　　D. addOnItemClickListener( )

2. 在使用 SimpleAdapter 时，需要将列表项数据组织成(   )。

  A. 数组　　　　　　　　　　　　　B. Set 集合

  C. List 集合　　　　　　　　　　　D. Map 集合

3. 在继承 BaseAdapter 实现自定义适配器时，实现将列表项数据放入列表项布局并返回列表项视图的方法是(   )。

  A. getCount( )　　　　　　　　　　B. getView( )

  C. getItem( )　　　　　　　　　　　D. getItemId( )

4. 下列关于 ListView 的说法中，正确的是(   )。

  A. ListView 的列表项不能设置单击事件

  B. ListView 不设置 Adapter 也能显示数据

  C. 每加载一条数据，适配器中的 getView( ) 方法都会被调用一次

  D. 当数据超出显示范围时，ListView 不能够自动滚动

5. 下面(   ) 方法不是自定义 Adapter 时需要重写的。

  A. getCount( )　　　　　　　　　　B. getView( )

  C. getItem( )　　　　　　　　　　　D. getItemPosition( )

二、技术实践

1. 利用 ListView 及 SimpleAdapter 完成如图 2-8 所示的好友列表效果。

2. 参考图 2-9、图 2-10、图 2-11 所示的页面，实现一个商品信息展示功能，单击商品图片则进入商品详情页面，单击"加入购物车"按钮，跳转至购物车页面。

图 2-8　好友列表效果图

图 2-9　商品列表页面

图 2-10　商品详情页面

图 2-11　购物车页面

任务 2　答题功能页面的设计与实现

任务描述

实现如图 2-12、图 2-13、图 2-14 所示的每日答题功能。每日答题总共 5 道题目，选择选项后单击"提交"按钮，将显示答案是否正确，并将"提交"按钮变为"下一题"按钮；单击"下一题"按钮切换题目；在答题结束后，弹出对话框，提示总得分。

任务学习目标

通过本任务需达到以下目标：
➢ 能够使用 RadioButton、CheckBox 实现单选及复选功能。
➢ 能够灵活运用 Fragment 进行页面开发。

图 2-12　单选题界面　　　图 2-13　提交后显示正误　　　图 2-14　多选题界面

技术储备

一、复选及单选按钮

（一）复选按钮（CheckBox）

复选按钮的使用

1. 作用

CheckBox 为复选按钮，用于实现多项选择功能。

2. 继承关系及常用属性

CheckBox 是抽象类 CompoundButton 的子类，Button 的间接子类。此类按钮均有两种状态：选中和未选中。当按钮被单击时，按钮选中状态自动发生变化。除了继承自 Button 的属性外，CheckBox 从 CompoundButton 继承了如下两个主要属性，如表 2-2 所示。

表 2-2　CheckBox 的主要属性

属性名称	功能描述
android：checked	设置按钮的勾选状态，true 表示勾选，false 表示未勾选，默认未勾选
android：button	设置左侧勾选图标的图形，如果不指定则使用系统的默认图标

注意：CheckBox 是 CompoundButton 一个最简单的实现。CompoundButton 分为两个部分，左侧的勾选图标和右侧的选项文字。如果将 android：button 值设为"@null"，则可以去掉左侧的勾选图标，如果不设置 android：text 的属性值，则仅有左侧的勾选图标，没有右侧的选项文字。

3. 事件处理

CheckBox 可以通过 setOnCheckedChangeListener 方法设置勾选状态变化监听器，监听器需要实现 CompoundButton.OnCheckedChangeListener 接口，具体代码如下。

```
1   //获取复选按钮对象
2   CheckBox checkbox = findViewById(R.id.checkbox);
3   checkbox.setOnCheckedChangeListener(
4                   new CompoundButton.OnCheckedChangeListener() {
5       @Override
6       public void onCheckedChanged(CompoundButton buttonView,bolean isChecked) {
7           //此方法第一个参数为勾选的按钮对象,第二个参数为选中状态
8           //事件处理代码
9       }
10  });
```

【案例 2-3】 编写代码实现如图 2-15 所示的界面效果，并实现单击选项时将选中的选项显示在题目后方。

1. 案例分析

界面中包含 3 个复选按钮，由于需要实现单击复选按钮将所选选项在题目后方显示或取消的功能，需要为每一个复选按钮添加 OnCheckedChangeListener 事件监听。

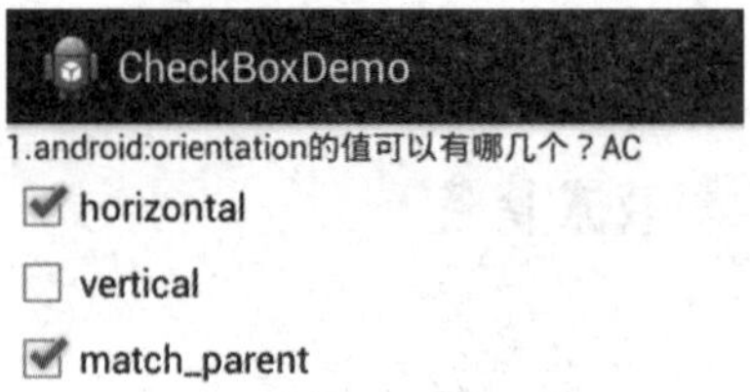

图 2-15 界面效果

2. 实现步骤

（1）创建名为"CheckBoxDemo"的程序。

（2）在 res/layout/activity_main.xml 中编写布局代码。

采用线性布局水平排列 4 个 TextView，1 个用于显示题干，另外 3 个在题干后面用于显示题目答案，垂直排列 3 个 CheckBox 用于显示选项。

```
1   <LinearLayout xmlns:android = "http://schemas.android.com/apk/res/android"
2       xmlns:tools = "http://schemas.android.com/tools"
3       android:layout_width = "match_parent"
4       android:layout_height = "match_parent"
5       android:orientation = "vertical"
6       tools:context = ".MainActivity" >
7       <LinearLayout
8           android:layout_width = "match_parent"
9           android:layout_height = "wrap_content" >
10          <TextView
11              android:layout_width = "wrap_content"
12              android:layout_height = "wrap_content"
13              android:text = "1.android:orientation 的值可以有哪几个?" />
14          <TextView
15              android:id = "@ + id/tv_answer1"
16              android:layout_width = "wrap_content"
```

```xml
17              android:layout_height = "wrap_content" />
18          <TextView
19              android:id = "@+id/tv_answer2"
20              android:layout_width = "wrap_content"
21              android:layout_height = "wrap_content" />
22          <TextView
23              android:id = "@+id/tv_answer3"
24              android:layout_width = "wrap_content"
25              android:layout_height = "wrap_content" />
26      </LinearLayout>
27      <CheckBox
28          android:id = "@+id/cb_choiceA"
29          android:layout_width = "match_parent"
30          android:layout_height = "wrap_content"
31          android:text = "horizontal" />
32      <CheckBox
33          android:id = "@+id/cb_choiceB"
34          android:layout_width = "match_parent"
35          android:layout_height = "wrap_content"
36          android:text = "vertical" />
37      <CheckBox
38          android:id = "@+id/cb_choiceC"
39          android:layout_width = "match_parent"
40          android:layout_height = "wrap_content"
41          android:text = "match_parent" />
42  </LinearLayout>
```

（3）在 MainActivity.java 中，编写功能逻辑代码。

由 MainActivity 实现 CompoundButton.OnCheckedChangeListener 接口，重写 onCheckedChanged( )方法，实现 CheckBox 控件的选项改变事件。同时，为 3 个 CheckBox 对象分别设置事件监听器。

```java
1  public class MainActivity extends AppCompatActivity implements
2                          CompoundButton.OnCheckedChangeListener {
3      private TextView tv_answer1,tv_answer2,tv_answer3;
4      private CheckBox cb_choiceA,cb_choiceB,cb_choiceC;
5      @Override
6      protected void onCreate(Bundle savedInstanceState) {
7          super.onCreate(savedInstanceState);
8          setContentView(R.layout.activity_main);
9          initView();
10     }
11     @Override
12     public void onCheckedChanged(CompoundButton buttonView,
13                                  boolean isChecked) {
```

```
14      switch(buttonView.getId()){
15          case R.id.cb_choiceA:
16              if(isChecked){
17                  tv_answer1.setText("A");
18              }else{
19                  tv_answer1.setText("");
20              }
21              break;
22          case R.id.cb_choiceB:
23              if(isChecked){
24                  tv_answer2.setText("B");
25              }else{
26                  tv_answer2.setText("");
27              }
28              break;
29          case R.id.cb_choiceC:
30              if(isChecked){
31                  tv_answer3.setText("C");
32              }else{
33                  tv_answer3.setText("");
34              }
35              break;
36      }
37  }
38  private void initView(){
39      //省略控件对象获取及添加事件监听语句
40  }
41 }
```

（二）单选按钮（RadioButton）

1. 作用

RadioButton 为单选按钮，用于实现在一组按钮中选择其中一项，其最大的特点在于选项间的互斥。

2. 继承关系及常用属性

RadioButton 的父类与 CheckBox 相同，也是 CompoundButton，其主要属性及特点与 CheckBox 基本相同，在此不再赘述。其显示效果如图 2-16 所示。

3. 使用方法

RadioButton 通常与单选组合框 RadioGroup 配合使用，以实现单选按钮间的互斥功能。RadioGroup 中可以包含多个 RadioButton。由于 RadioGroup 是 LinearLayout 的子类，可以通过 android：orientation 属性设置 RadioButton 的排列方式。

图 2-16 单选按钮

```
<RadioGroup
    ……
    android:orientation = "horizontal" >
  <RadioButton
    ……
    android:text = "选项1" />
……
</RadioGroup >
```

通过 RadioGroup 的 getCheckedRadioButtonId( )方法，可以获取当前选中的单选按钮 ID。

4. 事件处理

单选按钮的事件处理通常不由 RadioButton 处理，而是通过对 RadioGroup 设置事件监听实现，具体代码如下。

```
RadioGroup group = this.findViewById(R.id.radiogroup);
group.setOnCheckedChangeListener(new RadioGroup.OnCheckedChangeListener()
{
@Override
    public void onCheckedChanged(RadioGroup radioGroup,int checkedId) {
        //此方法第一个参数为 RadioGroup 对象,第二个参数为选中单选选项 ID
        //事件处理代码
    }
});
```

【案例2-4】 编写代码实现如图 2-17 所示的单选题界面效果，并将选中的答案显示在题干后。

1. 案例分析

界面中包含 TextView 控件，用于放题干及答案以及 1 个 RadioGroup，其包含 3 个 RadioButton，可以用相对布局进行控件布局。

MainActivity 中需要对 RadioGroup 添加事件监听，以实现单击选项即显示答案的功能。

图 2-17 界面效果

2. 实现步骤

（1）创建名为 "RadioButtonDemo" 的程序。

（2）在 res/layout/activity_main.xml 中编写布局代码。需注意：相对位置的参照控件应先定义。

```
1    <RelativeLayout xmlns:android = "http://schemas.android.com/apk/res/android"
2        xmlns:tools = "http://schemas.android.com/tools"
3        android:layout_width = "match_parent"
4        android:layout_height = "match_parent"
5        android:layout_margin = "10dp" >
6        <TextView
```

```xml
7          android:id = "@+id/question"
8          android:layout_width = "wrap_content"
9          android:layout_height = "wrap_content"
10         android:text = "1.RadioButton通常与(      )搭配使用。"/>
11     <TextView
12         android:id = "@+id/tv_answer"
13         android:layout_width = "wrap_content"
14         android:layout_height = "wrap_content"
15         android:layout_toRightOf = "@id/question"/>
16     <RadioGroup
17         android:id = "@+id/radiogroup"
18         android:layout_width = "match_parent"
19         android:layout_height = "wrap_content"
20         android:orientation = "horizontal"
21         android:layout_below = "@id/question" >
22         <RadioButton
23             android:id = "@+id/rb_optionA"
24             android:layout_width = "wrap_content"
25             android:layout_height = "wrap_content"
26             android:text = "Button" />
27         <!——此处省略选项B——>
28         <!——此处省略选项C——>
29     </RadioGroup>
30 </RelativeLayout>
```

（3）在MainActivity.java中编写代码，获取控件对象并为RadioGroup添加事件监听。

```java
1  public class MainActivity extends AppCompatActivity {
2  private TextView answer;
3  @Override
4  protected void onCreate(Bundle savedInstanceState) {
5      super.onCreate(savedInstanceState);
6      setContentView(R.layout.activity_main);
7      RadioGroup group = this.findViewById(R.id.radiogroup);
8      answer = this.findViewById(R.id.answer);
9      group.setOnCheckedChangeListener(
10                 new RadioGroup.OnCheckedChangeListener() {
11         @Override
12         public void onCheckedChanged(RadioGroup radioGroup,
13                                             int checkedId) {
14             //判断哪个单选按钮被选中
15             switch (checkedId){
16                 case R.id.rb_optionA:answer.setText("A");break;
17                 case R.id.rb_optionB:answer.setText("B");break;
```

```
18              case R.id.rb_optionC:answer.setText("C");break;
19          }
20        }
21    });
22  }
23 }
```

二、Fragment 碎片

（一）认识 Fragment

Fragment 是从 Android 3.0 开始推出的，起初目的是在大屏幕设备上实现 初识 Fragment
灵活、动态的界面设计，同时兼顾手机和平板的开发。如果界面中的控件比较多，界面布局比较复杂，可以用 Fragment 把屏幕划分成几个碎片，对界面进行模块化管理，让程序更加合理和充分地利用屏幕空间。

图 2-18、图 2-19 展示了同一个新闻应用界面在平板上和手机上的不同展示效果。由于平板屏幕较大，可以在一个 Activity 中包含左侧新闻列表和右侧新闻内容两个部分；而对于屏幕较小的手机，则需要在两个 Activity 中分别显示这两部分内容。为了能够更好地做到代码的复用，我们可以用一个 Fragment 来展示新闻列表，另一个 Framgent 用来展示新闻内容。

这里，Fragment 是一种嵌入 Activity 中的 UI 片段。一个 Activity 中可以包含多个 Fragment，一个 Fragment 也可以在多个 Activity 中使用。那 Fragment 应该如何创建和使用呢？

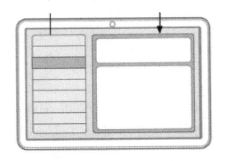  

图 2-18 新闻应用界面在平板上的效果　　图 2-19 新闻应用界面在手机上的效果

（二）创建 Fragment

Fragment 是独立的 UI 片段，必须写成可以重用的模块，即它有布局，也需要实现相应功能。因此，在创建一个 Fragment 的时候，需要同时创建布局文件和 Fragment 类，具体步骤如下。

在程序包名处右击，依次选择 New→Fragment→Fragment（Blank），进入图 2-20 所示的 Configure Component 界面，在该界面中指定 Fragment 名称以及对应的布局名称。

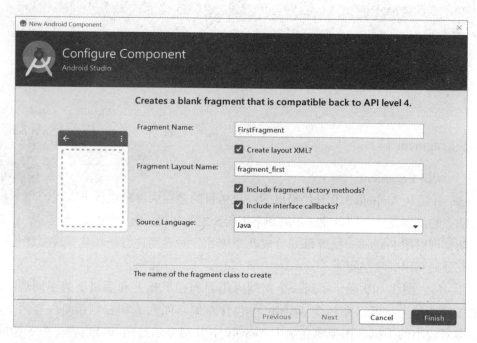

图 2-20 Configure Component 界面

创建完成后，会在 java 目录及 res/layout 目录中分别生成 Fragment 类和 Fragment 的布局文件。下面是生成的 FirstFragment 类的具体代码。

```
1    public class FirstFragment extends Fragment {
2        @Override
3        public View onCreateView(LayoutInflater inflater,ViewGroup container,Bundle savedInstanceState) {
4            return inflater.inflate(R.layout.fragment_first,container,false);
5        }
6    }
```

在上述代码中，FirstFragment 类继承自 Fragment，重写了 Fragment 的 onCreateView() 方法。在该方法中，通过 LayoutInflater 的 inflate() 方法将布局文件 fragment_first.xml 加载到 Fragment 中。

（三）添加 Fragment

Fragment 不能单独存在，必须添加到 Activity 中。向 Activity 中添加 Fragment 的方法通常有两种：一种是在布局文件中添加 Fragment，另一种是通过代码动态添加。

Fragment 基本用法

1. 在布局文件中添加 Fragment

如果将上面创建的 FirstFragment 添加到一个 Activity 中，只需要将 Fragment 作为一个控件加入 Activity 的布局文件即可。

```
1    <? xml version = "1.0" encoding = "utf-8"? >
2    <LinearLayout xmlns:android = "http://schemas.android.com/apk/res/android"
```

```
3      xmlns:app = "http://schemas.android.com/apk/res-auto"
4      xmlns:tools = "http://schemas.android.com/tools"
5      android:layout_width = "match_parent"
6      android:layout_height = "match_parent"
7      android:orientation = "horizontal"
8      tools:context = ".MainActivity" >
9  <fragment
10     android:id = "@+id/firstfragment"
11     android:layout_width = "match_parent"
12     android:layout_height = "wrap_content"
13     android:name = "com.example.helloworld.FirstFragment" />
14 </LinearLayout>
```

在上述代码中需注意：<fragment></fragment>标签的首字母需要小写，在标签中需要添加"android：name"属性，其属性值为 Fragment 的完整路径名。

2. 通过代码动态添加 Fragment

除了上述在布局文件中添加 Fragment 的方法外，还可以在 Activity 中动态添加 Fragment，具体步骤如下。

动态加载
Fragment

（1）创建一个 Framgent 对象。

（2）获取 FragmentManager 对象。

（3）开启 FragmentTransaction（Fragment 事务）。

（4）向 Activity 的布局中添加 Fragment。

（5）提交事务。

```
1  public class MainActivity extends AppCompatActivity {
2      @Override
3      protected void onCreate(Bundle savedInstanceState) {
4          super.onCreate(savedInstanceState);
5          setContentView(R.layout.activity_main);
6          //创建 Fragment 对象
7          FirstFragment firstFragment = new FirstFragment();
8          //获取 FragmentManager 对象
9          FragmentManager manager = getFragmentManager();
10         //获取 FragmentTransaction 对象
11         FragmentTransaction transaction = manager.beginTransaction();
12         //添加一个 Fragment
13         transaction.replace(R.id.layout,firstFragment);
14         //提交事务
15         transaction.commit();
16     }
17 }
```

上述代码中，需要特别注意特定方法的调用。getFragmentManager()方法获取 FragmentManager 对象；beginTransaction()方法获取 FragmentTransaction 对象；replace()方法添加

Fragment，需要两个参数，第 1 个参数为 Fragment 要加入的布局容器 ID，第 2 个参数为 Fragment 对象。Fragment 加入的布局容器一般是一个 FrameLayout，页面布局代码如下所示。

```xml
1  <?xml version = "1.0" encoding = "utf-8"?>
2  <LinearLayout xmlns:android = "http://schemas.android.com/apk/res/android"
3      xmlns:tools = "http://schemas.android.com/tools"
4      android:layout_width = "match_parent"
5      android:layout_height = "match_parent"
6      android:orientation = "vertical"
7      tools:context = ".MainActivity" >
8      <!-- 其他布局中的控件 -->
9      <FrameLayout
10         android:id = "@+id/layout"
11         android:layout_width = "match_parent"
12         android:layout_height = "match_parent" />
13     <!-- 其他布局中的控件 -->
14 </LinearLayout>
```

> **长知识：**
>
> **FrameLayout 布局**
>
> FrameLayout 又称帧布局，是最简单的布局。所有放在布局里的控件，都按照层次堆叠在屏幕的左上角，后加进来的控件覆盖前面的控件。控件可以通过 android:layout_gravity 属性控制自己在父控件中的位置。

3. Fragment 的生命周期

由于 Fragment 不能独立存在，因此 Fragment 的生命周期直接受所在 Activity 的影响。当在 Activity 中创建 Fragment 时，Fragment 处于启动状态；当 Activity 暂停时，其上的所有 Fragment 都暂停；当 Activity 被销毁时，其上的所有 Fragment 都被销毁。图 2-21 展示了 Fragment 的生命周期。

从图中可以看出，Fragment 的生命周期类似于 Activity。以下 5 个生命周期方法是 Fragment 生命周期独有的方法，具体说明如下。

- onAttatch()：Fragment 与 Activity 建立关联时调用。
- onCreateView()：创建 Fragment 视图时调用，主要用于加载布局；在该方法返回后，会立即调用 onViewCreate() 方法。
- onActivityCreated()：在 Activity 创建完毕后调用。
- onDestroyView()：Fragment 关联的视图被销毁时调用。
- onDetach()：Fragment 和 Activity 解除关联时调用。

【案例 2-5】 实现如图 2-18 和图 2-19 所示的新闻应用。横屏时显示效果如图 2-18 所示，单击左侧新闻列表，右侧更新新闻内容；竖屏时显示效果如图 2-19 所示，单击新闻列表跳转到新闻内容页面。

功能模块2　限时答题功能的实现

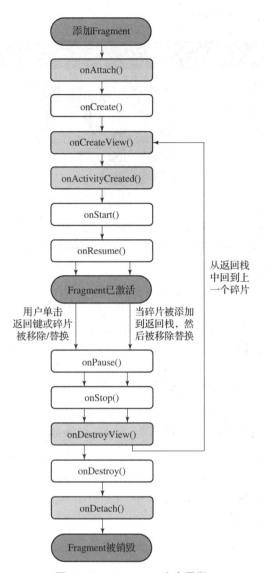

图 2 – 21　Fragment 生命周期

1. 案例分析

根据界面效果，可以用两个 Fragment 分别实现新闻列表及新闻内容。横屏效果下的 Activity 页面中包含这两个 Fragment，竖屏效果下，一个 Activity 中包含新闻列表 Fragment，另一个 Activity 中包含新闻内容 Fragment。具体界面分析及代码结构如图 2 – 22 所示。

2. 实现步骤

（1）创建名为"NewsDemo"的程序。

（2）创建 News 实体类。

News 代表新闻类，其中包含 title（标题）、content（内容）、image（新闻图片 id）3 个成员变量、构造方法及 getter、setter 方法，并实现序列化 Serializable 接口，以便能够通过 Intent 传递 News 对象。

实体类代码较为简单，此处省略。

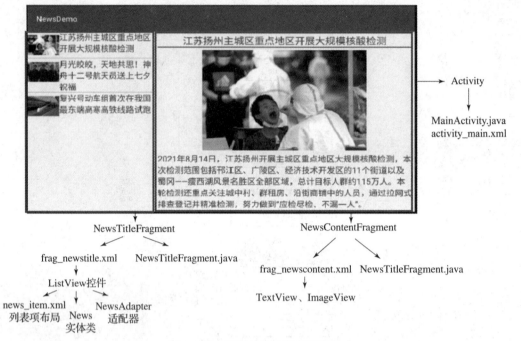

图 2-22 界面分析及代码结构图

（3）创建新闻列表 Fragment。

创建 NewsTitleFragment，对应布局文件为"frag_newstitle.xml"。新闻列表 Fragment 中包含新闻标题列表。在 frag_newstitle.xml 中，仅包含一个 ListView 控件即可。ListView 的实现需要创建列表项布局 news_item.xml 及自定义适配器 NewsAdapter.java。

由于 ListView 控件在 NewsTitleFragment 中，ListView 控件的功能逻辑代码应写在 NewsTitleFragment.java 中，具体代码如下。

```
1  public class NewsTitlesFragment extends Fragment {
2      private List <News >newsList ;
3      private NewsAdapter adapter;
4      private ListView titleList;
5      @Override
6      public View onCreateView(LayoutInflater inflater,ViewGroup container,
7      Bundle savedInstanceState) {
8          //初始化新闻数据
9          newsList = getNews();
10         View view = inflater.inflate(R.layout.news_title_frag,
11                                      container,false);
12         titleList =(ListView)view.findViewById(R.id.titlelist);
13         adapter = new NewsAdapter(getActivity(),R.layout.news_item,
14                                      newsList);
15         titleList.setAdapter(adapter);
16         return view;
```

```
17      }
18      //新闻数据初始化方法
19      private List <News >getNews(){
20          List <News >newsList = new ArrayList <News >();
21          News news1 = new News();
22          news1.setTitle("江苏扬州主城区重点地区开展大规模核酸检测");
23          news1.setContent("2021 年 8 月 14 日,江苏扬州开展主城区重点地区" +
24                  "大规模核酸检测,本次检测范围包括邗江区、广陵区、" +
25                  "经济技术开发区的 11 个街道以及蜀冈——瘦西湖风景名胜" +
26                  "区全部区域,总计目标人群约 115 万人。");
27          news1.setImage(R.drawable.new1);
28          //省略其他新闻数据初始化代码
29      }
30  }
```

（4）创建新闻内容 Fragment。

创建 NewsContentFragment，对应布局文件为"frag_newscontent.xml"。新闻内容 Fragment 布局中包含显示新闻标题（id：newstitle）、新闻内容（id：newscontent）及图片（id：newsimage）等控件。

在 NewsContentFragment.java 中，除了加载布局文件之外，还应完成与新闻内容相关的功能及操作。

> **问题解决 1：**
>
> **更新新闻内容谁来做？**
>
> 在程序设计中应做到各个部分各司其职。NewsContentFragment 是用来显示新闻的，那么单击新闻列表后，NewsContentFragment 中控件内容的更新应该由 NewsContentFragment 自己来做。因此，NewsContentFragment 中应该提供方法，单击新闻列表时将新闻标题和内容通过方法参数传递给 NewsContentFragment，由 NewsContentFragment 负责更新。

NewsContentFragment.java 具体代码如下。

```
1   public class NewsContentFragment extends Fragment {
2       private View view;
3       @Override
4       public View onCreateView(LayoutInflater inflater,ViewGroup container,
5       Bundle savedInstanceState) {
6           view = inflater.inflate(R.layout.news_content_frag,
7                                   container,false);
8           return view;
9       }
10      public void refresh(News news){
11          TextView content_newsTitle =
12                  (TextView)view.findViewById(R.id.newstitle);
```

```
13          content_newsTitle.setText(news.getTitle());
14          ImageView content_newImage =
15                  (ImageView)view.findViewById(R.id.newsimage);
16          content_newImage.setImageResource(news.getImage());
17          TextView content_newsContent =
18                  (TextView)view.findViewById(R.id.newscontent);
19          content_newsContent.setText(news.getContent());
20      }
21  }
```

在上述代码中，第 10~20 行定义了 refresh( ) 方法，通过 view 对象的 findVIewById( ) 方法获取控件对象，并将接收的 News 对象数据取出放入控件，实现新闻内容的更新。

（5）创建新闻内容页面 NewsContentActivity。

在竖屏情况下，新闻内容为一个单独的 Activity，其中包含 NewsContentFragment，通过选中新闻列表中某条新闻列表项跳转到 NewsContentActivity。

新闻内容页面的布局文件 activity_newscontent.xml 代码如下所示。

```
1   <?xml version="1.0" encoding="utf-8"?>
2   <LinearLayout xmlns:android="http://schemas.android.com/apk/res/android"
3       android:layout_width="match_parent"
4       android:layout_height="match_parent"
5       android:orientation="vertical" >
6       <fragment
7           android:id="@+id/news_content_fragment"
8           android:name="com.example.newsapp.NewsContentFragment"
9           android:layout_width="match_parent"
10          android:layout_height="match_parent" />
11  </LinearLayout>
```

问题解决 2：

如何实现单击新闻标题列表中的某条新闻，跳转到新闻内容 **Activity**，在新闻内容 **Activity** 中显示具体内容？

● 常规写法：创建 Intent 对象，将新闻信息放入 Intent，实现组件的通信。

● 改进写法：在 NewsContentActivity 中提供方法，NewsContentActivity 需要的数据定义为参数。这是一种启动活动的最佳方法，其原因是，在实际项目中，不同的 Activity 可能是由不同的程序员编写的，而编写 NewsContentActivity 的人最知道自己需要哪些参数，由它来定义接口，别的 Activity 不需要阅读 NewsContentActivity 的代码，只要按照参数来调用即可。

```
1   public class NewsContentActivity extends Activity {
2       @Override
3       protected void onCreate(Bundle savedInstanceState) {
```

```
4          super.onCreate(savedInstanceState);
5          setContentView(R.layout.news_content);
6          News news =(News)getIntent().getSerializableExtra("news");
7          NewsContentFragment newsContentFragment =
8                  (NewsContentFragment) getFragmentManager().
9                      findFragmentById(R.id.news_content_fragment);
10         newsContentFragment.refresh(news);
11      }
12      public static void actionStart(Context context,News news){
13         Intent intent =new Intent(context,NewsContentActivity.class);
14         intent.putExtra("news",news);
15         context.startActivity(intent);
16      }
17  }
```

在上述代码中，第 12～16 行定义了 actionStart( )方法，其参数为 Context 上下文对象和 News 对象。该方法将在竖屏情况下 NewsTitleFragment 中列表项被单击时被调用，跳转至 NewsContentActivity 页面。

第 6 行从 Intent 对象中取出从新闻列表页面中传递来的 News 对象；第 7～10 行代码获取 NewsContentFragment 对象，调用其 refresh( )方法进行新闻内容更新。

（6）在 activity_main.xml 中编写主界面布局代码。

在竖屏情况下，主界面中仅包含 NewsTitlesFragment；在横屏情况下，主界面中需要包含 NewsTitlesFragment 和 NewsContentFragment 两个部分。在横屏情况下，主界面布局代码如下所示。

```
1   <LinearLayout xmlns:android ="http://schemas.android.com/apk/res/android"
2       xmlns:tools ="http://schemas.android.com/tools"
3       android:layout_width ="match_parent"
4       android:layout_height ="match_parent"
5       tools:context =".MainActivity"
6       android:orientation ="horizontal" >
7       <!-- 新闻列表 Fragment -->
8       <fragment
9           android:id ="@+id/news_title_frag"
10          android:name ="com.example.newsapp.NewsTitlesFragment"
11          android:layout_weight ="1"
12          android:layout_width ="0dp"
13          android:layout_height ="wrap_content"
14          />
15      <!-- 新闻内容 Fragment -->
16      <FrameLayout
17          android:id ="@+id/news_content_layout"
18          android:layout_weight ="2"
19          android:layout_width ="0dp"
20          android:layout_height ="wrap_content" >
21          <fragment
```

```
22              android:id = "@ + id/news_content_frag"
23              android:name = "com.example.newsapp.NewsContentFragment"
24              android:layout_width = "match_parent"
25              android:layout_height = "wrap_content"
26              />
27      </FrameLayout >
28 </LinearLayout >
```

问题解决 3：

如何能够在一个应用程序中兼顾横屏和竖屏的不同界面显示效果？

思路：使用限定符。

在 res 目录下可以找到限定符的踪影，我们看到的 " – ldpi"" – mdpi"" – hdpi" 就是分辨率的限定符。除此之外，限定符还包括以下几种

（1）屏幕大小限定符：small、normal、large、xlarge。

（2）屏幕方向限定符：land、port。

（3）最小宽度限定符：sw600 dp。屏幕宽度大于 600 dp 就加载 layout – sw600 dp 下的布局，屏幕宽度小于这个值就加载默认 layout 下的布局。

在 res 目录下新建 "layout – land" 目录，将横屏的主界面布局放在此目录下，竖屏时的主界面布局文件仍保留在 layout 目录中，注意两个布局应该同名，且与 MainActivity 中加载的布局名相同。系统根据屏幕的方向，自动匹配布局文件。如果当前屏幕为横屏，则匹配 layout – land 目录中的布局文件，如果 layout – land 目录中不存在，则会匹配 layout 目录中的布局；如果当前屏幕为竖屏，则自动匹配 layout 目录中的布局文件。

（7）实现单击新闻列表显示新闻内容的功能。

竖屏情况与横屏情况下单击新闻标题项所做的操作是不同的，竖屏情况下跳转到内容页，横屏情况下直接在右侧的 Fragment 中更新内容。

问题解决 4：

如何在代码中区分当前是横屏还是竖屏？

思路 1：判断设备物理状态。

可以通过获取设备的配置信息 Configuration 对象，获取屏幕方向，进行判断，具体代码如下。

```
1 Configuration mConfig = getResources().getConfiguration();
2 int ori = mConfig.orientation;//获取屏幕方向
3 if(ori = = mConfig.ORIENTATION_LANDSCAPE){
4       //横屏
5 }else if(ori = = mConfig.ORIENTATION_PORTRAIT){
6       //竖屏
7 }
```

> **思路2**：判断布局中是否有右侧部分。
>
> 横、竖屏两种情况下，主界面布局的不同之处在于横屏时多了右侧的部分，可以用是否有右侧的部分作为判断两种情况的依据。

在 NewsTitleFragment 中，在布局完成加载、活动创建完毕后，在 onActivityCreated( )方法中，判断 activity_main.xml 布局中右侧的 FrameLayout 是否存在，具体代码如下。

```
1  public void onActivityCreated(Bundle savedInstanceState){
2      super.onActivityCreated(savedInstanceState);
3      if(getActivity().findViewById(R.id.news_content_layout)!=null)
4          isTwoPane=true;
5      else
6          isTwoPane=false;
7  }
```

在上述代码中，第3行判断右侧 FrameLayout 是否存在，其中，R.id.news_content_layout 即为横屏时主界面布局中右侧 FrameLayout 的 id；isTwoPane 为 NewsTitleFragment.java 中定义的布尔型变量，值为"true"时即为横屏状态。

新闻列表 ListView 控件的列表项单击事件处理中需根据 isTwoPane 的值进行不同操作，具体代码如下。

```
1  titleList.setOnItemClickListener(new OnItemClickListener(){
2      @Override
3      public void onItemClick(AdapterView<?>arg0,View arg1,
4                             int position,long arg3){
5          News news=newsList.get(position);
6          if(isTwoPane){
7              NewsContentFragment contentFragment=
8                      (NewsContentFragment)getFragmentManager().
9                      findFragmentById(R.id.news_content_frag);
10             contentFragment.refresh(news);
11         }else{
12             NewsContentActivity.actionStart(getActivity(),news);
13         }
14     }
15 });
```

在上述代码中，第7~10行为横屏情况下的情况，其中，第7~9行为右侧的 NewsContentFragment 对象，第10行调用 NewsContentFragment 对象的 refresh( )方法进行新闻内容的更新；第12行为竖屏情况下的情况，调用 NewsContentActivity 的 actionStart( )方法，跳转至 NewsContentActivity 页面进行新闻内容的显示。

◎ 任务实施

1. 任务分析

在本任务的限时答题功能中，有单选题和多选题两种题型，都包含题干、选项、答案、题目类型等属性,可以抽象出实体类 Question 类。

在界面显示时,需要根据题型的不同,切换单选题及复选题答题界面。可以将单选题和复选题答题界面分别用 Fragment 实现。题目的切换和题目答案的判断由 Fragment 负责。

在答题界面中,需要根据题型的不同,加载不同的 Fragment,并将题目对象传给 Fragment,接收 Fragment 返回的答题结果,并完成计算总分功能。

2. 实现步骤

在任务 1 中创建的答题界面(QuestionActivity)基础上,继续完成本任务功能。

(1) 在 entity 包中,创建实体类 Question。

【代码 2-6】 Question.java

```
1  public class Question{
2      private String questionStem;//题干
3      private String[] options;//选项
4      private String correctOptions;//正确选项的下标,从 0 开始
5      private int questionTypes;//1 为单选题,2 为多选题
6
7      //省略构造方法及 getter、setter 方法
8  }
```

(2) 在答题界面布局文件 activity_question.xml 中编写布局代码。

布局中包含用于显示倒计时的 TextView、"提交"按钮以及一个 FrameLayout,题目 Fragment 将添加在 FrameLayout 中,具体代码可扫描二维码查看。

代码:activity_question.xml

(3) 编写本任务所需的 shape 及 selector 选择器代码文件。

本任务中,单选、复选按钮选中及未选中时的按钮背景及文字显示效果不同,可以通过定义 shape 及 selector 选择器实现;同时,在提交答案后,选项会根据正确与否显示不同的背景及字体效果,也需要通过 shape 来实现。以下是本任务后续代码中需要用到的 shape 及 selector 的文件名及作用,如表 2-3 所示。具体代码此处不再列出,可以根据需要的效果自行设计。

表 2-3 shape 及 selector 的文件名及作用

文件名	作用	所在目录
selector_answer.xml	选项背景选择器,选项选中及未选中时显示不同效果	res/drawable
selector_answer_text.xml	选项文字选择器,选项选中及未选中时显示不同效果	res/color
shape_answer_select.xml	选项选中时的背景形状	res/drawable
shape_answer_not_select.xml	选项未选中时的背景形状	res/drawable
shape_answer_correct.xml	选项正确时的背景形状	res/drawable
shape_answer_fault.xml	选项错误时的背景形状	res/drawable

(4) 创建新的包 fragment，并在包中创建单选 Fragment 及复选 Fragment。

单选 Fragment 类为 SingleChoiceFragment，对应的布局文件为"fragment_singlechoice.xml"；复选 Fragment 类为 MultiChoiceFragment，对应的布局文件为"fragment_multichoice.xml"。

(5) 编写单选 Fragment 及复选 Fragment 的布局代码。

在单选 Fragment 的布局文件 fragment_singlechoice.xml 中编写布局代码，具体代码如【代码 2-7】所示。复选 Fragment 布局与单选 Fragment 基本类似，仅需将 RadioButton 控件更换为 CheckBox 控件，并去掉 RadioGroup 控件即可，具体代码可扫描二维码查看。

代码：fragment_multichoice.xml

【代码 2-7】 fragment_singlechoice.xml

```
1   <?xml version = "1.0" encoding = "utf-8"?>
2   <LinearLayout xmlns:android = "http://schemas.android.com/apk/res/android"
3       xmlns:tools = "http://schemas.android.com/tools"
4       android:layout_width = "match_parent"
5       android:layout_height = "match_parent"
6       android:orientation = "vertical"
7       android:padding = "20dp">
8       <!-- 头部 -->
9       <RelativeLayout
10          android:layout_width = "match_parent"
11          android:layout_height = "wrap_content">
12          <TextView
13              android:layout_width = "match_parent"
14              android:layout_height = "wrap_content"
15              android:text = "单选题"
16              android:textColor = "@color/black"
17              android:textSize = "22sp" />
18      </RelativeLayout>
19      <!-- 分割线 -->
20      <View
21          android:layout_width = "match_parent"
22          android:layout_height = "1dp"
23          android:layout_marginTop = "10dp"
24          android:background = "#bfbfbf" />
25      <!-- 题干 -->
26      <TextView
27          android:id = "@+id/question_stem"
28          android:layout_width = "match_parent"
29          android:layout_height = "wrap_content"
30          android:layout_marginTop = "20dp"
31          android:text = "题干"
32          android:textColor = "@color/black"
33          android:textSize = "20sp" />
34      <!-- 选项 -->
35      <RadioGroup
```

```
36          android:id = "@ + id/radioGroup"
37          android:layout_width = "match_parent"
38          android:layout_height = "wrap_content"
39          android:layout_marginTop = "20dp"
40          android:orientation = "vertical" >
41          <RadioButton
42              android:id = "@ + id/rb_1"
43              android:text = "选项1"
44              style = "@style/style_radiobutton" />
45          <!—省略其他选项-->
46      </RadioGroup>
47  </LinearLayout>
```

上述代码中,RadioButton 的属性设置中使用了样式 style_radiobutton,具体代码可扫描二维码查看。

(6) 在 SingleChoiceFragment. java 中,编写单选 Fragment 的功能逻辑代码。

代码:styles. xml

单选 Fragment 需要实现更新单选题目、判断答题结果的功能,因此在 Fragment 中,设计 nextQuestion( )方法用于更新题目,设计 getResult( )方法用于判断答题结果,具体代码如【代码 2 - 8】所示。

【代码 2 - 8】 SingleChoiceFragment. java

```
1   public class SingleChoiceFragment extends Fragment {
2       //成员变量
3       private TextView question_stem;    //题干文本框控件
4       private RadioGroup radioGroup;
5       private RadioButton[] rbs = new RadioButton[4];   //单选按钮控件
6       private Question question;       //题目对象
7       private int result;   //1:正确;0:错误;-1:未答题
8       @Override
9       public View onCreateView(LayoutInflater inflater,ViewGroup container,
10                                  Bundle savedInstanceState) {
11          return inflater.inflate(R.layout.fragment_singlechoice,
12                                  container,false);
13      }
14      @Override
15      public void onViewCreated(View view,Bundle savedInstanceState) {
16          super.onViewCreated(view,savedInstanceState);
17          initView(view);//初始化控件
18          nextQuestion(question);//更新题目
19      }
20      //获取题目
21      public void setQuestion(Question question) {
22          this.question = question;
23      }
24      //初始化控件
```

```java
25  private void initView(final View view) {
26      question_stem = view.findViewById(R.id.question_stem);
27      radioGroup = view.findViewById(R.id.radioGroup);
28      rbs[0] = view.findViewById(R.id.rb_1);
29      rbs[1] = view.findViewById(R.id.rb_2);
30      rbs[2] = view.findViewById(R.id.rb_3);
31      rbs[3] = view.findViewById(R.id.rb_4);
32  }
33  //更新题目
34  public void nextQuestion(Question question) {
35      //清空原选项
36      radioGroup.clearCheck();
37      //设置题干
38      question_stem.setText(question.getQuestionStem());
39      //设置选项
40      for (int i = 0; i < rbs.length; i++) {
41          rbs[i].setChecked(false);
42          rbs[i].setText(question.getOptions()[i]);
43          rbs[i].setBackgroundResource(R.drawable.selector_answer);
44          rbs[i].setTextColor(getResources().getColorStateList(
45                          R.color.selector_answer_text));
46      }
47  }
48  //判断答题结果,并返回 result
49  public int getResult(Question question) {
50      //正确选项的下标
51      int correctIndex = Integer.parseInt(question.getCorrectOptions());
52      int correctId = rbs[correctIndex].getId();   //正确选项单选按钮 ID
53      int checkedId = radioGroup.getCheckedRadioButtonId();
54      if(checkedId == -1){
55          Toast.makeText(getContext(),"请选择答案!",
56                      Toast.LENGTH_SHORT).show();
57          result = -1;
58      }else{
59          if (checkedId == correctId) {
60              //如果选中 ID 与正确选项 ID 相同则答案正确
61              result = 1;
62          } else{
63              //答案错误,选中项显示错误情况下的背景及文字颜色
64              RadioButton checkedButton =
65                          getView().findViewById(checkedId);
66              checkedButton.setBackground(getResources().getDrawable(
67                          R.drawable.shape_answer_fault));
68              checkedButton.setTextColor(getResources().getColor(
69                          R.color.answer_red_dark));
```

```
70              result = 0;
71          }
72          //无论答案是否正确,正确选项均显示正确情况下的背景及文字颜色
73          rbs[correctIndex].setBackgroundResource(
74                              R.drawable.shape_answer_correct);
75          rbs[correctIndex].setTextColor(getResources().getColor(
76                              R.color.answer_green_dark));
77      }
78      return result;
79  }
80 }
```

在上述代码中,第 34~47 行定义了 nextQuestion( ) 方法,其参数为 Question 对象,在方法中实现了清除上一次答题的选中项,设置题干、选项的功能;第 49~80 行定义了 getResult( ) 方法,其参数同样为 Question 对象,在方法中,获取了当前选中项的 ID 及正确选项的 ID,通过判断答题是否正确,对选项的背景及文字颜色进行设置。

(7) 在 MultiChoiceFragment. java 中,编写复选 Fragment 的功能逻辑代码。

复选 Fragment 的功能逻辑与单选 Fragment 基本一致,但由于复选按钮 (CheckBox) 与单选按钮 (RadioButton) 在使用上有所不同,在具体功能实现中的代码也略有不同,具体代码可扫描二维码查看。

代码:MultiCholce-Fragment. java

【代码 2-9】 MultiChoiceFragment. java 中的 getResult( ) 方法

```
1  public int getResult(Question question){
2      //得到正确的选项的下标数组
3      String[] correctOptions = question.getCorrectOptions().split(",");
4      boolean answer = true;//记录正确答案是否选择
5      int checkednum = 0;    //选择的选项个数
6      for(int i = 0;i < cbs.length;i + +){
7          if(cbs[i].isChecked())
8              checkednum + +;
9      }
10     if(checkednum = = 0){
11         Toast.makeText(getContext(),"请选择答案!",
12                         Toast.LENGTH_SHORT).show();
13         result = -1;
14     }else{
15         for(String correctOption : correctOptions) {
16             //判断对应的选项是否被选择
17             boolean now = cbs[Integer.parseInt(correctOption)].isChecked();
18             answer = answer && now;
19         }
20         if(answer){
21             //答案正确,将选项的背景及文字修改为正确状态
22             for (String correctOption : correctOptions) {
```

```
23              CheckBox cb = cbs[Integer.parseInt(correctOption)];
24              cb.setBackgroundResource(R.drawable.shape_answer_correct)
25              cb.setTextColor(getResources().getColor(
26                              R.color.answer_green_dark));
27          }
28          result =1;
29      }else{
30          //答案错误,将选项背景及文字颜色修改为错误状态
31          for (CheckBox checkBox : cbs) {
32              if (checkBox.isChecked()) {
33                  checkBox.setBackgroundResource(
34                              R.drawable.shape_answer_fault);
35                  checkBox.setTextColor(getResources().getColor(
36                              R.color.answer_red_dark));
37              }
38          }
39          showCorrectAnswer(correctOptions);//通过对话框显示正确结果
40          result = 0;
41      }
42  }
43  return result;
44 }
```

（8）在 QuestionActivity.java 中，编写功能逻辑代码。

在答题界面中，需要完成如下操作：

①答题结束，单击"提交"按钮，需要调用 Fragment 的 getResult( )方法判断答题结果，并根据返回结果，判断是否需要计分；同时，将按钮文本变为"下一题"；

②单击"下一题"按钮，需要根据当前题目的类型，加载单选或复选 Fragment；如果当前题目类型与前一题相同，则仅需调用 Fragment 的 nextQuestion( )方法，更新题目；否则，则需要在单选 Fragment 和复选 Fragment 之间进行切换。

流程图:ONCLICK
方法流程图

从上面的分析可以看出，功能逻辑的重点是按钮的单击事件处理，可以扫描二维码查看按钮单击事件的流程图，借助图示理解具体的功能逻辑。

QuestionActivity.java 的具体代码如【代码2-10】所示。

【代码2-10】　QuestionActivity.java

```
1  public class QuestionActivity1 extends AppCompatActivity{
2      private int score =0;  //答题得分
3      private final List<Question> questions = new ArrayList<>();//题目集合
4      private int i =0;//当前题号
5      private int previousType;//上一题题目类型,单选:1,多选:2
6      private int result;  //答题结果 1:正确;0:错误;-1:未答题
7      private Button btn_submit_next;//提交/下一题按钮
8      private SingleChoiceFragment singleFragment;   //单选题 Fragment 对象
9      private MultiChoiceFragment multiFragment;     //多选题 Fragment 对象
```

```java
10      @Override
11      protected void onCreate(Bundle savedInstanceState) {
12          super.onCreate(savedInstanceState);
13          setContentView(R.layout.activity_answer1);
14          //初始化题目
15          initQuestions();
16          //初始化控件
17          initView();
18          //根据题型加载第1题
19          Question question = questions.get(0);
20          if(question.getQuestionTypes() = =1){
21              singleFragment.setQuestion(question);
22              replaceFragment(singleFragment);
23              previousType =1;    //当前题型为1,单选题
24          }else{
25              multiFragment.setQuestion(question);
26              replaceFragment(multiFragment);
27              previousType =2;    //当前题型为2,多选题
28          }
29      }
30      private void initView() {
31          //找到对应的控件
32          btn_submit_next = findViewById(R.id.btn_submit_next);
33          singleFragment = new SingleChoiceFragment();
34          multiFragment = new MultiChoiceFragment();
35          //设置单击事件监听
36          btn_submit_next.setOnClickListener(new View.OnClickListener() {
37              @Override
38              public void onClick(View v) {
39                  //如果是"提交"按钮,根据当前题目类型,调用getResult()
40                  if (btn_submit_next.getText().equals("提交") ) {
41                      if (questions.get(i).getQuestionTypes() = =1)
42                          result = singleFragment.getResult(questions.get(i));
43                      else
44                          result =multiFragment.getResult(questions.get(i));
45                      //如果答对了,则得分加1
46                      if (result = =1)
47                          score + +;
48                      //如果答题了,且当前不是第5题则按钮文本改为"下一题"
49                      //如果当前是第5题,则文本改为"结束"
50                      if(i = =4)
51                          btn_submit_next.setText("结束");
52                      else if(result ! = -1)
53                          btn_submit_next.setText("下一题");
54                  } else {
```

```
55              //如果是"下一题"按钮,题号增加,判断是否已经答题完成
56              i++;
57              if(i<questions.size()){
58                  //根据题目类型及上一题类型,选择是否替换Fragment
59                  if(questions.get(i).getQuestionTypes()==1){
60                      if(previousType!=1){
61                          singleFragment.setQuestion(questions.get(i));
62                          replaceFragment(singleFragment);
63                          previousType=1;
64                      }else
65                          singleFragment.nextQuestion(questions.get(i));
66                  }else{
67                      if(previousType!=2){
68                          multiFragment.setQuestion(questions.get(i));
69                          replaceFragment(multiFragment);
70                          previousType=2;
71                      }else
72                          multiFragment.nextQuestion(questions.get(i));
73                  }
74                  btn_submit_next.setText("提交");
75              }else{
76                  showDialog();
77                  btn_submit_next.setVisibility(View.INVISIBLE);
78              }
79          }
80      }
81  });
82  }
83  //替换Fragment方法,当第一次加载题目或题型发生变化时调用该方法
84  private void replaceFragment(Fragment fragment){
85      FragmentManager fragmentManager=getSupportFragmentManager();
86      FragmentTransaction transaction=fragmentManager.beginTransaction();
87      transaction.replace(R.id.answer_con,fragment);
88      transaction.commit();
89  }
90  private void initQuestions(){
91      //第1题
92      Question question1=new Question();
93      question1.setQuestionStem("第十四届全国冬季运动会第一次"+
94                              "在冬运会上设置_____季竞技项目。");
95      question1.setOptions(new String[]{"春","夏","秋","冬"});
96      question1.setCorrectOptions("1");
97      question1.setQuestionTypes(1);
98      questions.add(question1);
99      //第2题
```

```
100        Question question2 = new Question();
101        question2.setQuestionStem("参加中国共产党第一次全国代表大会" +
102                         "的上海的代表是_____和_____。");
103        question2.setOptions(new String[]{"李达,李汉俊","董必武,陈潭秋",
104                         "邓恩铭,王尽美","毛泽东,周恩来"});
105        question2.setCorrectOptions("0");
106        question2.setQuestionTypes(1);
107        questions.add(question2);
108        //其余题目省略
109    }
110    //显示答题结束的dialog
111    private void showDialog() {
112        AlertDialog.Builder builder = new AlertDialog.
113                         Builder(QuestionActivity1.this);
114     builder.setTitle("提示")
115         .setMessage("答题结束,得分为" + score)
116         .setPositiveButton("确定",
117           new DialogInterface.OnClickListener() {
118             @Override
119             public void onClick(DialogInterface dialog,int which) {
120                 Intent intent = new Intent(QuestionActivity1.this,
121                         PointsActivity.class);
122                 startActivity(intent);
123                 finish();
124             }
125          });
126     builder.create().show();
127    }
128 }
```

任务反思

编写并运行程序，将在代码编写及程序调试过程中出现的异常信息、产生原因及解决方法记录在下方。

问题1：_____

产生原因：_____

解决方法：_____

问题2：_____

产生原因：_____

_____

解决方法：_____

任务总结及巩固

师兄：在这个任务中，我们学习的一个重点内容就是Fragment，你能对Fragment做一个总结吗？

小白：Fragment就像报纸上的一个小专栏，有内容、有功能、也有自己的生命周期，几个Fragment组成一个Activity，也可以将一个Fragment复用到多个不同的地方。有了Fragment就可以更加灵活和方便地对界面进行组织了。

师兄：总结得不错，在使用Fragment的时候，还有一点需要格外注意，就是Fragment要"扛得起责任"，简单说就是自己的控件要自己管。不能出现控件在Fragment的布局文件中，具体与控件相关的功能逻辑却还写在Activity里的情况。也就是说，在开发过程中，要让各个模块职责明确。我们作为开发人员，在工作中，也要扛起责任。往小了说，我们交出去的代码要经得起检验，软件的质量保证有我们一份；往大了说，中国软件产业的崛起也有我们一份。

小白：听你这么一说，还真是责任重大，我得好好学好技术！这些知识和技术本身感觉并不难，但是在具体的项目实践中，可真是问题不断。在这个任务的实现过程中，可真是花了不少时间呢！

师兄：那可不是嘛！每个功能实现起来细节都不少，有的逻辑也挺复杂。咱们一定要做到思路清晰、逻辑理顺、代码规范，才能保质保量完成项目开发。这可不是一天两天的事情，需要不断磨炼和学习！给你个自学任务：RadioButton和CheckBox还有一个"兄弟"叫Switch——开关按钮，回去查查资料，看看Switch怎么使用吧！

一、基础巩固

1. 实现单选按钮（RadioButton）的互斥功能，需要使用（   ）控件。

   A. CheckBox   B. LinearLayout   C. RadioGroup   D. ViewGroup

2. 设置CheckBox选择事件监听通常使用（   ）方法。

   A. setOnClickListener（）   B. setOnItemClickListener（）

   C. setOnItemSelectedListener（）   D. setOnCheckedChangeListener（）

3. 如果需要去掉单选按钮和复选按钮左侧的图标，可以进行（   ）设置。

   A. android：button = @null   B. android：text = @null

C. android：background = @null D. android：src = @null

4. 下列关于 Fragment 的说法，错误的是（   ）。

  A. Fragment 可以更加合理地进行界面布局

  B. Fragment 需要在 Manifest.xml 文件中声明

  C. Fragment 不能够独立存在，需要嵌入 Activity 中

  D. 一个 Activity 中可以有多个 Fragment

5. 下列方法中，（   ）在 Fragment 创建视图时调用。

  A. onAttach( ) B. onCreateView( )

  C. onActivityCreate( ) D. onDestroyView( )

二、技术实践

利用本任务中学习的内容自行设计一个长征中重要历史事件回顾程序，需要完成不少于 3 个历史事件的展示，单击选择事件名称时，需要显示事件的详细信息及图片。

任务3 使用数据库存储题目信息

任务描述

创建数据库存储题目信息，实现从数据库中随机读取 5 道题目用于每日限时答题。

任务学习目标

通过本任务需达到以下目标：

➢ 能够创建 SQLite 数据库并实现数据的增、删、改、查功能。

➢ 能够使用内容提供者（ContentProvider）共享数据。

技术储备

一、使用 SQLite 数据库存储数据

对于大量数据的存储，数据库当仁不让。在 Android 系统中提供了 SQLite 数据库用于数据存储。SQLite 是一种轻量级关系型数据库，运算速度快，占用资源少，特别适合在移动设备上使用。SQLite 数据库支持标准的 SQL 语法，但比一般的数据库简单得多，不需要安装、不需要设置用户名和密码就可以使用。

（一）SQLite 数据库的创建

下面我们要创建一个名为"UserManager.db"的数据库，然后在这个数据库中创建一张"users"表，用来存放用户信息。

要创建 SQLite 数据库，需要先创建一个类继承 SQLiteOpenHelper 类，在该类中重写 onCreate( )方法和 onUpgrade( )方法，具体代码如下所示。

数据库的创建

```
1   class MyDataBaseHelper extends SQLiteOpenHelper {
2
3       public static final String CREATE_USERS = "create table users ("
4               + "id integer primary key autoincrement,"
5               + "username text,"
6               + "password text,"
7               + "age integer)";
8       private Context mContext;
9       public MyDatabaseHelper(Context context,String name,
10                      SQLiteDatabase.CursorFactory factory,int version){
11          super(context,name,factory,version);
12          mContext = context;
13      }
14      @Override
15      public void onCreate(SQLiteDatabase db) {
16          db.execSQL(CREATE_USERS);
17          Toast.makeText(mContext,"数据库创建成功!",
18                                  Toast.LENGTH_SHORT).show();
19      }
20      @Override
21      public void onUpgrade(SQLiteDatabase db,int oldVersion,int newVersion){
22      }
23  }
```

在上述代码中，第 3~7 行将建表语句定义为一个字符串常量，接下来定义了 3 个方法。第 9~14 行定义了构造方法，该方法需要 4 个参数。第 1 个参数是 Context 上下文对象；第 2 个参数是数据库名；第 3 个参数允许我们在查询数据时返回一个自定义的 Cursor，一般传入"null"；第 4 个参数表示当前数据库的版本号。第 15~19 行的 onCreate() 方法中，在第 16 行通过调用 execSQL() 方法执行建表语句。

此时，并没有创建数据库。我们还需要创建 MyDataBaseHelper 对象，调用对象的 getReadableDatabase() 或 getWritableDatabase() 方法，才能创建数据库，具体代码如下。

```
1   public class MainActivity extends AppCompatActivity {
2
3       private MyDataBaseHelper dbHelper;
4       @Override
5       protected void onCreate(Bundle savedInstanceState) {
6           super.onCreate(savedInstanceState);
7           setContentView(R.layout.activity_main);
8           dbHelper = new MyDataBaseHelper(this,"UserManager.db",null,1);
9           dbHelper.getWritableDatabase();
10      }
11  }
```

在上述代码中，第 8 行创建了一个 MyDataBaseHelper 对象，通过参数指定数据库名为

"UserManager.db",版本号为1;第9行调用了 getWritableDatabase() 方法创建数据库。当第一次运行程序时,系统会检测到当前程序中没有 UserManager.db 数据库,就会创建数据库并调用 MyDataBaseHelper 中的 onCreate() 方法,users 表被创建,并弹出提示信息"数据库创建成功!"。当关闭程序、再次打开时,会发现不再有提示信息弹出,这是因为系统检测到数据库已经存在,不再执行数据库及数据表创建语句。

（二）SQLite 数据库的升级

在项目开发中及产品上线之后,随着软件产品需求的改变、功能的完善,我们会遇到数据库需要跟随需求而改变的情况,比如,需要在 users 表中新增一列手机号码或者需要新增部门表。

数据库的升级

下面我们模拟在原来 users 表的基础上,再增加部门表 department,具体建表语句如下所示。

```
1  create table department(
2      id integer primary key autoincrement,
3      departmentname text,
4      departmentcode text
5  )
```

下面采用3种方法进行数据库的更新。

方法1：在 onCreate() 方法中,执行 department 建表语句。

```
1  class MyDataBaseHelper extends SQLiteOpenHelper {
2
3  public static final String CREATE_USERS = "create table users ("
4          + "id integer primary key autoincrement,"
5          + "username text,"
6          + "password text,"
7          + "age integer)";
8  public static final String CREATE_DEPARTMENT = "create table department ("
9          + "id integer primary key autoincrement,"
10         + " departmentname text,"
11         + " departmentcode text)";
12     private Context mContext;
13
14     public MyDatabaseHelper(Context context,String name,
15                 SQLiteDatabase.CursorFactory factory,int version){
16         super(context,name,factory,version);
17         mContext = context;
18     }
19     @Override
20     public void onCreate(SQLiteDatabase db) {
21         db.execSQL(CREATE_USERS);
22         db.execSQL(CREATE_DEPARTMENT);
```

```
23            Toast.makeText(mContext,"数据库创建成功!",
24                    Toast.LENGTH_SHORT).show();
25        }
26        @Override
27        public void onUpgrade(SQLiteDatabase db,int oldVersion,int newVersion){
28        }
29  }
```

重新运行程序,我们会发现并没有弹出"数据库创建成功!"的提示信息,并没有创建 department 表。这是由于只要程序已经安装过,数据库就已经存在了,MyDataBaseHelper 中的 onCreate() 方法就不会再次执行。除非将程序卸载,数据库被删除,onCreate() 方法才会再次执行,department 表才会被创建。

但每次数据库有更新都卸载程序显然是不合理的,这就需要通过 SQLiteOpenHelper 中提供的升级方法进行数据库的更新。

方法 2:在 onUpgrade() 方法中,进行数据库的更新。

下面代码在 onUpgrade() 方法中加入了数据库更新代码。

```
1  class MyDataBaseHelper extends SQLiteOpenHelper {
2    //省略原代码
3    @Override
4    public void onUpgrade(SQLiteDatabase db,int oldVersion,int newVersion){
5        db.execSQL("drop table if exists users");
6        db.execSQL("drop table if exists department");
7        onCreate(db);
8    }
9  }
```

在 onUpgrade() 方法中,首先执行了两条 drop 语句,如果 users 表和 department 表存在,则将两个表删除,然后调用 onCreate() 方法重新创建数据表。这种方法可以避免因数据表存在重新创建时导致程序报错。

要想让 onUpgrade() 方法执行,需要在创建 MyDataBaseHelper 对象时,在构造方法的第 4 个参数中传入当前数据库的版本号。最初创建 MyDataBaseHelper 对象时传入的版本号是 1,现在只需传入比 1 大的数,就可以让 onUpgrade() 方法执行,具体代码如下。

```
1  public class MainActivity extends AppCompatActivity {
2
3      private MyDataBaseHelper dbHelper;
4      @Override
5      protected void onCreate(Bundle savedInstanceState) {
6          super.onCreate(savedInstanceState);
7          setContentView(R.layout.activity_main);
8          dbHelper=new MyDataBaseHelper(this,"UserManager.db",null,2);
9          dbHelper.getWritableDatabase();
10     }
11 }
```

在上述代码的第 8 行传入的版本号为 2，onUpgrade( )方法执行，删除原数据表 users 并重新创建两个数据表。

这种方法将原数据表删除，这是一个非常危险的操作，会使原程序中存储的本地数据全部丢失，那是否有更好的方法进行数据库升级呢？

方法 3：根据版本号进行数据库的升级。

每一个新的数据库版本都会对应一个版本号，当指定的数据库版本号大于当前数据库版本号时，就会进入 onUpgrade( )方法执行更新操作。可以为每一个版本号设置它需要更改的内容，在 onUpgrade( )方法中对当前数据库版本号进行判断，再执行相应的数据库更新操作。

```
1   class MyDataBaseHelper extends SQLiteOpenHelper {
2       //省略原代码
3       @Override
4       public void onUpgrade(SQLiteDatabase db,int oldVersion,int newVersion){
5           switch(oldVersion){
6               case 1: db.execSQL(CREATE_DEPARTMENT);
7               case 2: ……
8           }
9       }
10  }
```

在 onUpgrade( )方法中，通过 switch 判断当前的版本号是 1，就会只创建 department 表；如果还有第 3 个版本的数据库更新，就可以在 switch 中添加新的 case 语句，判断当前版本号是否为 2。这里需要注意，case 后没有 break 语句，这样可以保证在跨版本升级时，每一次数据库的修改语句都能被全部执行。

（三）SQLite 数据库的查看

前面已经创建了数据库，但数据库是否创建成功呢？又该如何查看数据库中的数据呢？数据库的存储位置与 SharedPreferences 相同，在 data/data/包名/目录下。在创建数据库时会在该目录下创建 databases 目录，数据库文件就在这个目录中。但数据库创建完成后是无法直接对数据进行查看的，要想查看数据可以使用 Android 调试桥。

数据库的查看

Android 调试桥（adb）是 Android SDK 中自带的调试工具，通过它可以与设备进行通信。它存放在 SDK 的 platform – tools 目录下，可以通过命令行窗口打开 adb，具体步骤如下。

步骤 1：配置环境变量。

将 SDK 的 platform – tools 目录添加到环境变量中系统变量的 Path 变量中。

步骤 2：打开命令行窗口，输入"adb shell"命令，即可进入设备的控制台，如图 2 – 23 所示。

步骤 3：使用 cd 命令，转到 data/data/应用名/databases 目录下，如图 2 – 24 所示。

图 2-23 "adb shell" 命令

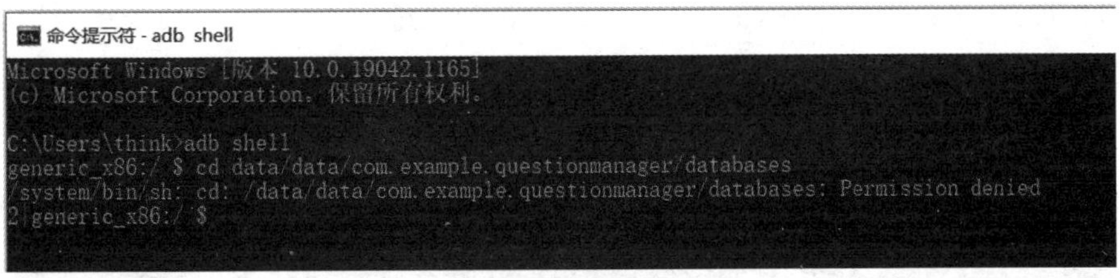

图 2-24 cd 命令

如果此时窗口中提示"Permission denied",则表示没有查看系统文件的权限,需要通过"su root"命令,更换系统用户,如图 2-25 所示。

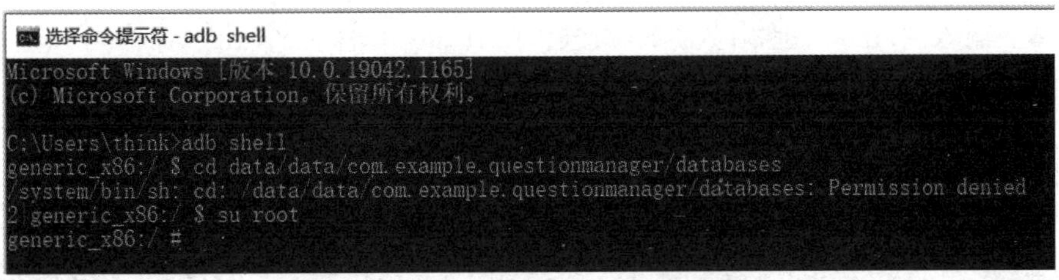

图 2-25 "su root" 命令

> **注意**:如果模拟器的系统镜像为带有 Google Play 的版本,则无法通过"su root"命令获取系统用户权限。

步骤 4:使用 ls 命令,查看该目录下的文件,如图 2-26 所示。

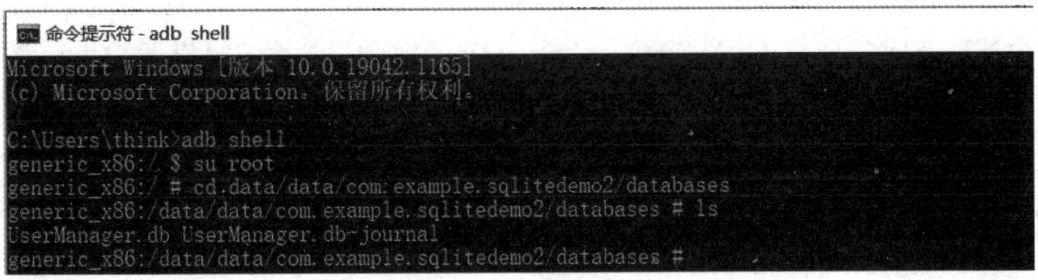

图 2-26 ls 命令

从图 2-26 中可以看出,databases 目录下有两个数据库文件,一个是创建的 UserManager.db,另一个是临时日志文件 UserManager.db-journal。

步骤 5:输入"sqlite3 数据库名",打开数据库,如图 2-27 所示。

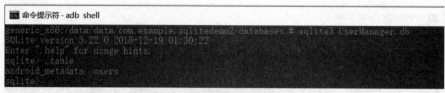

图 2-27　打开数据库

输入".table"命令,可以看到此时数据库中有两张表,users 表是前面创建的数据表,android_metadata 表是每个数据库都会自动生成的表。通过".schema"命令还可以查看建表语句,也可以通过 sql 语句进行数据库操作。具体如图 2-28 所示。

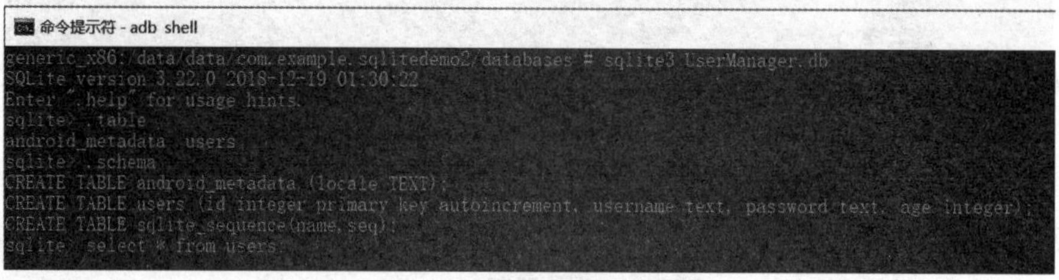

图 2-28　查看及操作数据库

步骤 6:输入".quit"退出数据库调试,回到 adb;输入"exit"命令可以退出设备控制台,如图 2-29 所示。

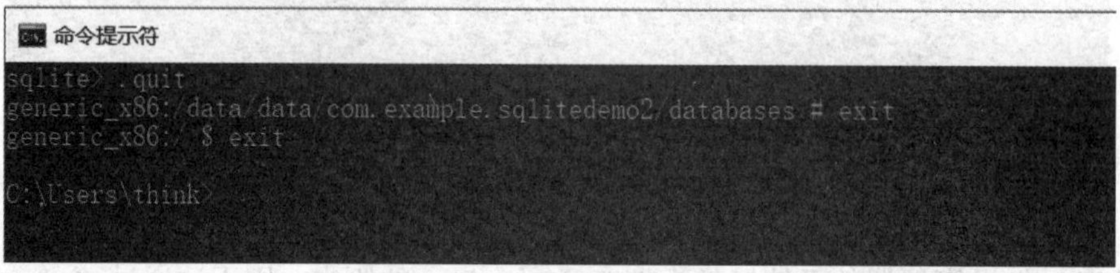

图 2-29　退出设备控制台

(四) SQLite 数据库的基本操作

数据库创建完毕,下面对数据库进行数据的增、删、改、查操作。在创建数据库时,我们调用了 MyDataBaseSQLiteHelper 对象的 getWritableDatabase() 或 getReadableDatabase() 方法,这两个方法都会返回一个 SQLiteDatabase 对象。SQLiteDatabase 类中提供了 SQLite 数据库操作的各种方法,通过 SQLiteDatabase 对象就可以对数据进行增、删、改、查等基本操作。

1. 添加数据

SQLiteDatabase 类中提供了 insert() 方法,实现向数据表中插入一条数据,insert() 方法的定义如下所示。

添加数据

```
1  public long insert(String table,String nullColumnHack,
2                                  ContentValues values)
```

insert() 方法需要 3 个参数,第 1 个参数为数据表的名称,第 2 个参数表示在未指定添加数据的情况下将某些可以为空的列自动复制为 null,第 3 个参数为 ContentValues 对象,它提供了一系列的 put( ) 方法向 ContentValues 中添加数据,需要将表中的列名及对应的数据作为参数传递给 put( ) 方法。

insert( ) 方法执行完成后,如果插入成功则返回行号,否则返回 "-1"。

向 users 数据表中添加一条用户信息的具体代码如下所示。

```
1   public long insertUser(String username,String password,int age){
2       MyDataBaseHelper myHelper = new MyDataBaseHelper(context,
3                           "UserManager.db",null,1);
4       SQLiteDatabase db = myHelper.getWritableDatabase();
5       ContentValues values = new ContentValues();
6       values.put("username",username);
7       values.put("password",password);
8       values.put("age",age);
9       long id = db.insert("users",null,values);
10      db.close();
11      return id;
12  }
```

在上述代码中,定义了 insertUser( ) 方法进行数据的添加,将用户名 username、密码 password、年龄 age 作为参数传入方法;第 4 行获取 SQLiteDatabase 对象;第 5 行创建了 ContentValues 对象;第 6~8 行将用户数据通过 put( ) 方法放入 values 对象,其中第 1 个参数为数据表中的列名,第 2 个参数为对应的数据;第 9 行调用 SQLiteDatabase 对象的 insert( ) 方法添加数据。

需要注意的是,使用完 SQLiteDatabase 对象后一定要像第 10 行一样调用 close( ) 方法关闭数据库连接,否则数据库连接会一直存在,不断消耗内存,当系统内存不足时将获取不到 SQLiteDatabase 对象,并会报出数据库未关闭的异常。

2. 删除数据

SQLiteDatabase 类中提供了 delete( ) 方法实现删除数据表中数据的功能,delete( ) 方法的定义如下所示。

```
1   public int delete (String table,String whereClause,String[] whereArgs)
```

delete( ) 方法需要 3 个参数,第 1 个参数为数据表的名称,第 2 个参数为删除的条件,其中可以用 "?" 作为参数占位符,第 3 个参数为第 2 个参数中占位符所需值组成的字符串数组。

删除 users 数据表中某条用户数据的具体代码如下所示。

```
1   public int deleteUser(long id){
2       MyDataBaseHelper myHelper = new MyDataBaseHelper(context,
3                   "UserManager.db",null,1);
4       SQLiteDatabase db = myHelper.getWritableDatabase();
5       int num = db.delete("users","id = ?",new String[]{id + ""});
6       db.close();
7       return num;
8   }
```

在上述代码中，定义了 deleteUser( )方法，根据传入的 id 进行数据的删除；第 5 行调用 SQLiteDatabase 对象的 delete( )方法，以 "id 等于传入的 id 值" 作为删除条件，方法返回删除数据的条数。

删除和更新数据

3. 修改数据

SQLiteDatabase 类中提供了 update( )方法实现数据表中数据的更新，update( )方法的定义如下所示。

```
public int update(String table,ContentValues values,String whereClause,String[] whereArgs)
```

update( )方法需要 4 个参数，第 1 个参数为数据表的名称，第 2 个参数为一个 ContentValues 对象，用于存放新数据，第 3 个参数表示要修改的数据的查询条件，第 4 个参数为查找条件。

修改 users 表中某个用户的密码和年龄，具体代码如下所示。

```
1   public int updateUser(String username,String password,int age){
2       MyDataBaseHelper myHelper = new MyDataBaseHelper(context,
3                                    "UserManager.db",null,1);
4       SQLiteDatabase db = myHelper.getWritableDatabase();
5       ContentValues values = new ContentValues();
6       values.put("password",password);
7       values.put("age",age);
8       int number = db.update("users",values,"name = ?",
9                                    new String[]{username});
10      db.close();
11      return number;
12  }
```

在上述代码中，定义了 updateUser( )方法进行数据的修改，将用户名为 username 的用户密码改为 "password"、年龄改为 "age"；第 5~7 行创建了 ContentValues 对象，放入需要修改的数据；第 8~9 行将调用 SQLiteDatabase 对象的 update( )方法修改数据。

4. 查询数据

SQLiteDatabase 类中提供了 query( )方法实现数据的查询，query( )有多个重载方法，其中，参数最少的、也是我们最常用的方法的定义如下所示。

```
public Cursor query (String table,String[]columns,String selection,String[] selectionArgs,String groupBy,String having,String orderBy)
```

该方法需要 7 个参数，分别为表名、要查询的列名数组、查询条件、查询条件中的参数数组、分组方式、having 条件、排序方式。这些参数与 SQL 中的 select 查询语句中的各部分相互对应，具体对应情况见表 2-4。该方法返回一个 Cursor 对象，Cursor 是一个游标接口，提供了便利查询结果的方法。

查询数据

表 2-4 参数对应情况

方法参数	对应 SQL 部分	含义
table	from table_name	查询的表名
columns	select column1，column2	查询的列名
selection	where column = value	Where 查询条件
selectionArgs	—	为 where 中的占位符提供具体值
groupBy	groupby column	需要分组的列
having	having column = value	对分组后的结果进一步约束
orderBy	orderby column1，column2	排序方式

查询数据的具体代码如下所示。

```
1  publicArrayList queryAll(){
2     ArrayList list = new ArrayList();
3     MyDataBaseHelper myHelper = new MyDataBaseHelper(context,
4                                 "UserManager.db",null,1);
5     SQLiteDatabase db = myHelper.getWritableDatabase();
6     Cursor cursor = db.query("information",null,null,null,null,
7                                 null,null);
8     list = convertFromCursor(cursor);
9     return list;
10 }
```

其中，convertFromCursor( )方法通过 Cursor 对象遍历查询结果，并将其返回为一个 List 集合，具体代码如下所示。

```
1  private ArrayList convertFromCursor(Cursor cursor){
2     ArrayList list = new ArrayList();
3     if(cursor! = null && cursor.moveToFirst()){
4     //通过游标遍历整个查询结果集
5     do{
6     int id = cursor.getInt(cursor.getColumnIndex("id"));
7     String username = cursor.getString(cursor.getColumnIndex("username"));
8     String password = cursor.getString(cursor.getColumnIndex("password"));
9      int age = cursor.getInt(cursor.getColumnIndex("age"));
10     User user = new User(id,username,password,age);
11     list.add(user);
12    }while(cursor.moveToNext());
13    }
14    return list;
15 }
```

在上述方法中，第 2 行创建了 ArrayList 对象，用于存放查询到的用户数据；第 3 行判断

cursor 是否为空、是否查询到数据，其中，调用 cursor 对象的 moveToFirst()方法，将 cursor 移到查询的第一条记录，如果有查询结果，该方法返回"true"，否则，返回"false"。

第 5~13 行循环读取查询结果中的每条记录，并通过 moveToNext()方法将光标移到下一条记录，直到该方法返回"false"，所有记录读取完成；在读取记录时，调用了 getColumnIndex()方法传入列名，获取该列的列索引，并调用 getString()、getInt()方法传入列索引，返回该列对应的数据；将读取的数据封装为 User 对象（需创建包含 id、username、password、age 成员变量的实体类 User），加入 list 集合。

【案例 2-6】 完成一个具有增、删、改、查功能的用户管理页面，界面效果如图 2-30 所示。

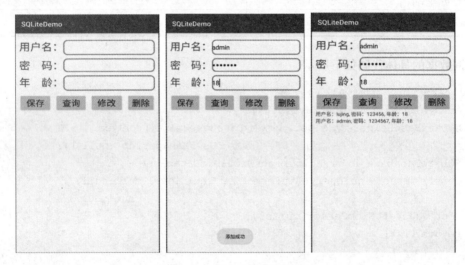

图 2-30　用户管理界面效果

1. 案例分析

前面已经创建了用户管理数据库及数据表，并学习了对数据进行增、删、改、查的方法。本案例中，将数据操作方法封装为一个 UserDao 类，在 MainActivity 中，调用 UserDao 类中的方法实现数据操作，并对返回结果进行处理。

2. 实现步骤

（1）创建名为"SQLiteDemo"的应用程序。

（2）在 activity_main.xml 布局文件中完成界面布局。

布局中包括 4 个 TextView 控件，分别用于显示用户名、密码、年龄及查询结果，包括 3 个 EditText 控件，分别用于输入用户名、密码和年龄，包括 4 个 Button 控件，分别用于添加、查询、修改及删除数据。

（3）创建 UserDao 类，封装增、删、改、查方法。

在 UserDao 类的构造方法中，创建 MyDataBaseHelper 对象，代码如下所示。

```
1  public class UsersDao {
2      private MyDataBaseHelper myHelper;
3
4      public UsersDao(Context context) {
5          myHelper = new MyDataBaseHelper(context,
6                              "UserManager.db",null,1);
7      }
8      //添加数据
9      public long insertUser(String username,String password,int age) {
10         //具体代码省略
11     }
12     //删除数据
13     public int deleteUser(String username) {
14         //具体代码省略
15     }
16     //更新数据
17     public int updateUser(String username,String password,int age) {
18         //具体代码省略
19     }
20     //查询数据
21     public ArrayList  queryAll (){
22         //具体代码省略
23     }
24     //将查询数据组织为集合
25     private ArrayList convertFromCursor(Cursor cursor) {
26         //具体代码省略
27     }
28 }
```

（4）在 MainActivity 中编写逻辑代码，实现用户信息的添加、查询、修改以及删除功能。

由于需要为"保存""查询""修改""删除"按钮设置单击事件，MainActivity 需实现 OnClickListener 接口，并重写 onClick( )方法，具体代码如下。

```
1  public class MainActivity extends AppCompatActivity implements
2                               View.OnClickListener {
3      private UsersDao dao;
4      private EditText et_username,et_password,et_age;
5      private Button bt_save,bt_query,bt_update,bt_delete;
6      private TextView tv_show;
7      @Override
8      protected void onCreate(Bundle savedInstanceState) {
9          super.onCreate(savedInstanceState);
10         setContentView(R.layout.activity_main);
11         dao = new UsersDao(this);
```

```java
12        init();
13    }
14    //控件及变量初始化
15    public void init(){
16        //代码省略
17    }
18    @Override
19    public void onClick(View v) {
20        switch (v.getId()){
21            case R.id.bt_save: {     //添加数据
22                String username = et_username.getText().toString();
23                String password = et_password.getText().toString();
24                int age = Integer.parseInt(et_age.getText().toString());
25                long id = dao.insertUser(username,password,age);
26                if (id ! = -1) {
27                    Toast.makeText(this,"添加成功",
28                                    Toast.LENGTH_LONG).show();
29                } else {
30                    Toast.makeText(this,"添加失败",
31                                    Toast.LENGTH_LONG).show();
32                }
33                break;
34            }
35            case R.id.bt_query: {    //查询数据
36                ArrayList users = dao.queryAll();
37                if (users.size() = =0) {
38                    tv_show.setText("没有数据");
39                } else {
40                    for (int i =0;i<users.size();i + +) {
41                        User user = (User) users.get(i);
42                        tv_show.append("用户名:"+user.getUsername() +
43                                    ",密码:"+user.getPassword() +
44                                    ",年龄:"+user.getAge() +" \n");
45                    }
46                }
47                break;
48            }
49            case R.id.bt_update: {   //更新数据
50                String username = et_username.getText().toString();
51                String password = et_password.getText().toString();
52                int age = Integer.parseInt(et_age.getText().toString());
53                int num = dao.updateUser(username,password,age);
54                if(num ! =0){
55                    Toast.makeText(this,"修改成功",
56                                    Toast.LENGTH_LONG).show();
57                } else {
58                    Toast.makeText(this,"修改失败",
```

```
59                                        Toast.LENGTH_LONG).show();
60                  }
61              break;
62              }
63          case R.id.bt_delete: {    //删除数据
64              String username = et_username.getText().toString();
65              int num = dao.deleteUser(username);
66              if(num! =0){
67                  Toast.makeText(this,"删除成功",
68                                        Toast.LENGTH_LONG).show();
69              } else {
70                  Toast.makeText(this,"删除失败",
71                                        Toast.LENGTH_LONG).show();
72              }
73              break;
74          }
75       }
76    }
77 }
```

在上述代码中，第21~34行获取界面上输入的用户信息，调用UserDao类的insertUser()方法将用户信息添加到数据库中。

第35~48行调用UserDao类的queryAll()方法获得User对象集合users，遍历集合取出User对象，将用户信息显示在界面上。

第49~62行获取界面上输入的用户信息，调用UserDao类的updateUser()方法将用户密码和年龄进行更新。

第63~74行调用UserDao类的deleteUser()方法将用户名为"username"的数据从数据库中删除。

长知识：

使用SQL语句进行数据库操作

除了用上述介绍的方法进行数据库操作之外，还可以使用execSQL()方法，通过SQL语句对数据库进行增、删、改、查操作，具体方法如下。

```
1  //增加一条数据
2  db.execSQL("insert into users(username,price) values (?,?)",
3                              new Object[]{name,price});
4  //删除一条数据
5  db.execSQL("delete from users where id =1");
6  //修改一条数据
7  db.execSQL("update users set password =? where username =?",
8                              new Object[]{password,username});
9  //查询数据
10 Cursor cursor = db.rawQuery("select * from userswhere name =?",
11                              new String[]{name});
```

二、使用内容提供者共享数据

在我们使用移动应用程序时,经常会遇到这样的情况:当打开支付宝的"添加朋友"功能时,会推荐可能认识的人,并详细地标明哪些是手机联系人。支付宝是如何获取我们的手机联系人的呢?这就需要联系人程序将它保存的联系人信息共享给支付宝等应用程序。为了实现跨程序共享数据的需求,Android 系统提供了内容提供者,它能实现安全可靠的数据共享。

(一) 认识内容提供者

内容提供者是 Android 四大组件之一,它提供了在不同应用程序之间实现数据共享的完整机制,允许一个程序访问另一个程序中的数据。内容提供者可以选择只提供哪些数据的共享,从而保证了隐私数据的安全性。使用内容提供者是 Android 实现跨程序数据共享的标准方式。图 2-31 展示了内容提供者的工作原理。

在图 2-31 中,A 程序通过 ContentProvider 将数据共享,B 程序需要通过 ContentResolver 操作 A 程序共享的数据,A 程序将操作结果返回给 ContentResolver,ContentResolver 将操作结果返回给 B 程序。

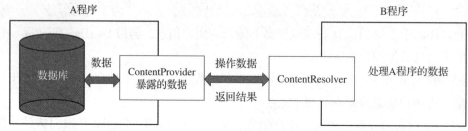

图 2-31 内容提供者工作原理图

(二) 使用 ContentResolver 访问共享数据

当一个应用程序通过 ContentProvider 提供了对外访问的接口,任何其他的应用程序就都可以对共享的数据进行访问。在前面我们说到联系人程序将联系人信息共享给支付宝等应用程序,就是通过 ContentProvider 来实现的。除了联系人程序,Android 系统中自带的短信、媒体库等程序也提供了类似的访问接口。那支付宝是如何访问联系人共享的数据的呢?

要想访问 ContentProvider 中共享的数据,需要借助 ContentResolver 类,ContentResolver 实例可以通过 Context 的 getContextResolver() 方法获取。ContentResolver 中提供了一系列方法对数据进行增、删、改、查操作,insert() 方法用于添加数据,update() 方法用于更新数据,delete() 方法用于删除数据,query() 方法用于查询数据,方法的具体定义形式见表 2-5。

表 2-5 方法的具体定义

方法定义	参数含义
Uri insert (Uri uri, ContentValues value)	Uri:内容 URI;value:插入数据
int update (Uri uri, ContentValues value, String where, String [] whereArgs)	Uri:内容 URI;value:更新数据 where:where 约束条件 whereArgs:where 中占位符的具体值

续表

方法定义	参数含义
int delete（Uri uri, String where, String []whereArgs）	Uri：内容 URI；where：where 约束条件 whereArgs：where 中占位符的具体值
Cursor query（Uri uri, String []projection, String selection, String []selectionArgs, String sortOrder）	Uri：内容 URI；projection：查询的列名 selection：where 约束条件 selectionArgs：where 中占位符的具体值 sortOrder：查询结果的排序方式

这些方法与前面在数据库中学习的 SQLiteDatabase 中的方法非常类似，不同之处在于，ContentResolver 中的增、删、改、查方法的第 1 个参数不再是表名，而是一个内容 URI。内容 URI 是 ContentProvider 中数据的唯一标识，它主要由 3 个部分组成：协议（sheme）、权限（authority）和路径（path）。图 2-32 是一个 URI 的组成结构图。

协议是以"content：//"开头的前缀，不需要修改；权限用于对不同的应用程序作区分，为了避免冲突，一般会采用程序包名来命名；路径则是用于对同一应用程序中不同的表作区分，上图中的 users 则表示 users 数据表。通过 URI 可以清楚地区分要访问的是哪个应用程序中哪张表的数据。在作为参数传入上述方法时，需要将内容 URI 字符串通过 Uri 的 parse()方法解析成 Uri 对象，代码如下所示。

```
Uri uri = Uri.parse("content://com.example.sqlitedemo/users")
```

以上 4 个方法的用法与 SQLiteDatabase 中的方法基本相同，这里不再赘述。下面通过【案例 2-7】来具体实践。

【案例 2-7】 读取系统联系人中的联系人信息，如图 2-33 所示，并显示在如图 2-34 所示的列表中。

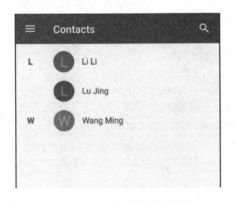

图 2-33 联系人程序中的列表

图 2-34 读取的联系人列表

1. 案例分析

Anroid 系统中自带的联系人、短信等应用程序已经定义了 ContentProvider 共享数据接口。本案例中，需要获取 ContentResolver 对象，通过内容 URI 确定要访问的数据。联系人应用的内容 URI 为常量"ContactsContract. CommonDataKinds. Phone. CONTENT_URI"，在调用 ContentResolver 对象的 query( )方法进行联系人信息查询时，需要将常量作为第 1 个参数。

值得注意的是，在 Android 6.0 及以上版本中，读取联系人信息属于危险权限，需要进行静态权限声明及动态权限申请。

2. 实现步骤

（1）创建名为"ContentDemo"的应用程序。

（2）在 activity_main. xml 中，编写布局代码。

显示联系人需要使用 ListView 控件，在布局代码中加入一个 ID 为"listview"的 ListView 控件。代码较为简单，此处不再列出。

（3）在 AndroidManifest. xml 文件中，添加读取联系人信息权限声明。

```xml
<uses-permission android:name="android.permission.READ_CONTACTS"/>
```

（4）在 MainActivity 中，编写功能逻辑代码。

```java
1  public class MainActivity extends AppCompatActivity {
2      private ListView listview;
3      private ArrayAdapter adapter;
4      private ArrayList<String> list = null;
5      @Override
6      protected void onCreate(Bundle savedInstanceState) {
7          super.onCreate(savedInstanceState);
8          setContentView(R.layout.activity_main);
9          //控件初始化
10         listview = findViewById(R.id.listview);
11         list = new ArrayList();
12         adapter = new ArrayAdapter(this,
13                     android.R.layout.simple_list_item_1,list);
14         listview.setAdapter(adapter);
15         //获取权限
16         if (ContextCompat.checkSelfPermission(this,
17                     Manifest.permission.READ_CONTACTS)
18                         != PackageManager.PERMISSION_GRANTED) {
19             ActivityCompat.requestPermissions(this,
20                 new String[]{Manifest.permission.READ_CONTACTS},1);
21         }else{
22             readContact();
23         }
24     }
25     @Override
```

```java
26    public void onRequestPermissionsResult(int requestCode,
27                        String[] permissions,int[] grantResults){
28        super.onRequestPermissionsResult(requestCode,permissions,
29                                        grantResults);
30        if (requestCode = =1) {
31            if (grantResults ! =null &&
32 grantResults[0] = = PackageManager.PERMISSION_GRANTED) {
33                readContact();
34            }
35        }
36    }
37    //读取联系人信息
38    public void readContact(){
39        Uri uri = ContactsContract.CommonDataKinds.Phone.CONTENT_URI;
40        //1.获取 ContentResolver 对象
41        ContentResolver resolver = getContentResolver();
42        //2.查询联系人数据
43        Cursor cursor = resolver.query(uri,null,null,null,null);
44        //3.通过游标遍历数据
45        list = convertCursor(cursor);
46        //4.通知 adapter 数据集发生变化,需要刷新
47        adapter.notifyDataSetChanged();
48        //5.关闭游标
49        cursor.close();
50    }
51    //将查询结果转换为集合
52    public ArrayList<String> convertCursor(Cursor cursor){
53        String name = ContactsContract.CommonDataKinds.Phone.DISPLAY_NAME;
54        String phone = ContactsContract.CommonDataKinds.Phone.NUMBER;
55        if (cursor ! = null && cursor.moveToFirst()) {
56            do {
57                String contactName = cursor.getString(
58                                cursor.getColumnIndex(name));
59                String contactNum = cursor.getString(
60                                cursor.getColumnIndex(phone));
61                list.add(contactName + " \n" + contactNum);
62            } while (cursor.moveToNext());
63        }
64        return list;
65    }
66 }
```

在上述代码中,第 10~14 行,在 onCreate( )方法中获取 ListView 控件对象,配置适配器对象;第 16~23 行,动态申请读取联系人权限,该权限为常量"Manifest. permission. READ_CONTACTS",如果没有获得权限,则弹出权限申请对话框,否则,调用 readContact( )方法。

第26~36行为动态权限申请回调方法,当用户授予权限时,同样调用readContact( )方法获取联系人数据。

第38~49行定义了readContact( )方法,在该方法中获取ContentResolver对象,调用query( )方法查询共享数据,并将返回的Cursor对象转换为集合赋值给全局变量list;第47行代码调用了adapter适配器对象的notifyDataSetChanged( )方法,通知适配器数据集合list发生变化,刷新列表,显示更新后的数据。

第52~66行定义了convertCursor( )方法,将Cursor对象转换为集合,在获取联系人姓名及电话时,姓名列的列名对应常量"ContactsContract"。电话列的列名对应常量"ContactsContract"。获取到的数据直接添加到全局变量list中。

(三) 创建自己的内容提供者

前面通过读取联系人信息,我们学习了如何访问其他应用程序的共享数据。那应用程序是如何通过内容提供者实现数据共享的呢?如果我们想要把自己应用程序中的数据共享给其他应用程序,又该如何做呢?下面通过3个步骤创建自己的内容提供者。

步骤1: 创建一个新类继承ContentProvider。

ContentProvider是一个抽象类,其中有6个抽象方法。继承ContentProvider后,需要重写6个抽象方法。

```java
1   public class MyProvider extends ContentProvider {
2       @Override
3       public boolean onCreate() {
4           return false;
5       }
6       @Override
7       public Cursor query(Uri uri,String[]projection,String selection,
8                           String[]selectionArgs,String sortOrder) {
9           return null;
10      }
11      @Override
12      public String getType(Uri uri) {
13          return null;
14      }
15      @Override
16      public Uri insert(Uri uri,ContentValues contentValues) {
17          return null;
18      }
19      @Override
20      public int delete(Uri uri,Stringselection,String[] selectionArgs) {
21          return 0;
22      }
23      @Override
24      public int update(Uri uri,ContentValues contentValues,
25                        Stringselection,String[] selectionArgs) {
26          return 0;
27      }
28  }
```

从这 6 个方法的定义中可以看到，其中 query( )、insert( )、update( )、delete( ) 与前面在 ContentResolver 中调用的方法完全相同，通过 ContentResolver 对象调用的增、删、改、查方法与这里定义的方法相对应，真正完成数据增、删、改、查的是在 ContentProvider 中定义的方法。

onCreate( ) 方法在初始化 ContentProvider 时调用，通常会将数据库的创建和升级等操作放在该方法中；该方法返回"true"表示 ContentProvider 初始化成功，返回"false"表示初始化失败。只有当 ContentResolver 尝试访问共享数据时，ContentProvider 才会被初始化。

getType( ) 方法根据传入的内容 URI 返回相应的 MIME 类型。

步骤 2：实现匹配内容 URI 功能。

上述 6 个方法中除 onCreate( ) 方法外，其余 5 个方法都需要接收 URI 参数。ContentProvider 中的一个重要的功能就是要实现 URI 的解析，分析出外部程序需要访问的表和数据。

前面我们已经了解内容 URI "content：//com. example. sqlitedemo/users" 表示外部程序希望访问 com. example. sqlitedemo 应用的 users 表中的数据。除了这种标准写法外，还可以进一步精确到表中的某条数据。内容 URI "content：//com. example. sqlitedemo/users/2" 就表示外部程序希望访问 com. example. sqlitedemo 应用的 users 表中 id 为 2 的数据。也可以使用通配符来匹配这两种格式的内容 URI，具体规则如下。

- *表示匹配任意长度的任意字符。
- #表示匹配任意长度的数字。

一个能匹配任意表的内容 URI 可以写成：

content：//com. example. sqlitedemo/ *

一个能匹配 users 表中的任意一行数据的内容 URI 可以写成：

content：//com. example. sqlitedemo/users/#

下面通过 UriMatcher 类实现内容 URI 的匹配功能，我们在 MyProvider 类中加入如下代码。

```
1   public class MyProvider extends ContentProvider {
2       public static final int USERS_DIR = 0;
3       public static final int USERS_ITEM = 1;
4       private static UriMatcher matcher;
5
6       static {
7           matcher = new UriMatcher(UriMatcher.NO_MATCH);
8           matcher.addURI("com.example.sqlitedemo2","users",USERS_DIR);
9           matcher.addURI("com.example.sqlitedemo2","users/#",USERS_ITEM);
10      }
11      @Override
12      public Cursor query(Uri uri,String[]projection,String selection,
13                          String[]selectionArgs,String sortOrder){
14          switch (matcher.match(uri)){
```

```
15          case USERS_DIR:
16              //查询 users 表中的所有数据
17              break;
18          case USERS_ITEM:
19              //查询 users 表中的单条数据
20              break;
21      }
22      return null;
23   }
24   //省略其他方法代码
25 }
```

在上述代码中,第 2、3 行定义了两个整型常量,其中"USERS_DIR"表示访问 users 表中的所有数据,"USERS_ITEM"表示访问 users 表中的单条数据。

第 6~11 行的静态代码块中,创建 UriMatcher 的实例,调用 addURI( ) 方法将期望匹配的内容 URI 格式传递进去。该方法接收 3 个参数,分别为权限、路径和前面自定义的常量,其中,路径参数可以使用通配符。

第 13~24 行的 query( ) 方法中,通过 UriMatcher 的 match( ) 方法对传入的 Uri 对象进行匹配,如果匹配成功,match( ) 方法返回相应的自定义常量,就可以根据判断结果明确外部程序希望访问的是哪些数据,即可执行相应的数据库操作。

步骤 3:完善 getType( ) 方法。

getType( ) 方法是所有 ContentProvider 必须提供的方法,用于获取 Uri 对象对应的 MIME 类型。一个内容 URI 对应的 MIME 字符串主要由 3 个部分组成,其具体要求如下。

- 必须以"vnd"开头。
- 如果内容 URI 以路径结尾,则后面接"android.cursor.dir/";如果内容 URI 以 id 结尾,则后面接"android.cursor.item/"。
- 最后加上"vnd.<authority>.<path>"

按照上述规则,内容 URI "content://com.example.sqlitedemo/users" 的 MIME 类型为:
vnd.android.cursor.dir/vnd.com.example.sqlitedemo.users

内容 URI "content://com.example.sqlitedemo/users/2" 的 MIME 类型为:
vnd.android.cursor.item/vnd.com.example.sqlitedemo.users

MyProvider 类中 getType( ) 方法可以进行如下完善。

```
1 public class MyProvider extends ContentProvider {
2     //省略其他方法代码
3     @Override
4     public String getType(@NonNull Uri uri) {
5         switch (matcher.match(uri)){
6             case USERS_DIR:
7                 return "vnd.android.cursor.dir/vnd.com.example.
8                                              sqlitedemo.users";
```

```
9           case USERS_ITEM:
10              return "vnd.android.cursor.item/vnd.com.example.
11                                                  sqlitedemo.users";
12          default:
13              break;
14      }
15      return null;
16  }
17 }
```

【案例 2-8】 继续完善 MyProvider,实现案例【2-6】中的用户管理数据共享,并创建应用程序对用户管理数据进行增、删、改、查操作。

1. 案例分析

通过前面 3 个步骤,已经基本完成了 MyProvider,现需要在增、删、改、查 4 个方法中添加具体的数据库操作语句。同时,在 AndroidManifest.xml 文件中对 MyProvider 进行声明。

在另一个应用程序中,仍然需要通过 ContentResolver 调用增、删、改、查方法实现对用户管理数据的操作。

2. 实现步骤

(1) 在【案例 2-6】项目中,创建 MyProvider 类。

在前面完成的 MyProvider 代码基础上,在 onCreate() 方法中创建 MyDataBaseHelper 对象,创建数据库。注意,MyDataBaseHelper 中使用的 Toast 需要去掉,在跨程序访问数据时不能使用 Toast。同时,对增、删、改、查方法进行完善。

```
1  public class MyProvider extends ContentProvider {
2
3      public static final int USERS_DIR = 0;
4      public static final int USERS_ITEM = 1;
5      public static final String AUTHORITY = "com.example.sqlitedemo2";
6      private static UriMatcher matcher;
7      private MyDataBaseHelper helper;
8      //省略静态代码块
9      //省略 getType()方法
10     @Override
11     public boolean onCreate() {
12         helper = new MyDataBaseHelper(getContext(),"UserManager.db",
13                                                      null,1);
14         return true;
15     }
16     //查询方法
17     @Override
18     public Cursor query(Uri uri,String[] projection,String selection,
19                    String[] selectionArgs,String sortOrder) {
```

```
20          SQLiteDatabase db = helper.getReadableDatabase();
21          Cursor cursor = null;
22          switch (matcher.match(uri)){
23              case USERS_DIR:
24                  //查询 users 表中的所有数据
25                  cursor = db.query("users",projection,selection,
26                                  selectionArgs,null,null,sortOrder);
27                  break;
28              case USERS_ITEM:
29                  //查询 users 表中的单条数据
30                  String userId = uri.getPathSegments().get(1);
31                  cursor = db.query("users",projection,"id = ?",
32                                  new String[]{userId},null,null,sortOrder);
33                  break;
34              default:
35                  break;
36          }
37          return cursor;
38      }
39      //插入方法
40      @Override
41      public Uri insert(Uri uri,ContentValues contentValues) {
42          SQLiteDatabase db = helper.getReadableDatabase();
43          Uri returnUri = null;
44          switch (matcher.match(uri)){
45              case USERS_DIR:
46              case USERS_ITEM:
47                  //添加数据不区分全表及单条数据
48                  long id = db.insert("users",null,contentValues);
49                  returnUri = Uri.parse("content://" + AUTHORITY + "/users/" + id);
50                  break;
51              default:
52                  break;
53          }
54          return returnUri;
55      }
56      //删除方法
57      @Override
58      public int delete(Uri uri,String selection,String[] selectionArgs) {
59          SQLiteDatabase db = helper.getReadableDatabase();
60          int deleteRows = 0;
61          switch (matcher.match(uri)){
62              case USERS_DIR:
63                  deleteRows = db.delete("users",selection,selectionArgs);
64                  break;
```

```
65            case USERS_ITEM:
66                //根据id删除数据
67                String userId = uri.getPathSegments().get(1);
68                deleteRows = db.delete("users","id = ?",new String[]{userId});
69                break;
70            default:
71                break;
72        }
73        return deleteRows;
74    }
75    //更新方法
76    @Override
77    public int update(Uri uri,ContentValues contentValues,String selection,
78    String[] selectionArgs) {
79        SQLiteDatabase db = helper.getReadableDatabase();
80        int updateRows = 0;
81        switch (matcher.match(uri)){
82            case USERS_DIR:
83                updateRows = db.update("users",contentValues,selection,
84                                                selectionArgs);
85                break;
86            case USERS_ITEM:
87                //根据id修改数据
88                String userId = uri.getPathSegments().get(1);
89                updateRows = db.update("users",contentValues,"id = ?",
90                                                new String[]{userId});
91                break;
92            default:
93                break;
94        }
95        return updateRows;
96    }
97 }
```

在上述代码中，第 16~37 行重写了 query( ) 方法，方法中先获取了 SQLiteDatabase 实例，然后根据传入的 Uri 判断用户想要访问哪张表，是进行整表查询还是单条数据的查询；接下来调用 SQLiteDatabase 的 query( ) 进行查询，并将 Cursor 对象返回。在访问单条数据时，第 30 行代码调用了 Uri 对象的 getPathSegments( ) 方法，将内容 URI 权限之后的部分以 "/" 进行分割，并把分割后的结果放入一个字符串列表中，列表中第 0 个位置上存放的是路径，第 1 个位置上存放的是 ID。取出 ID 后，通过设置 selection 和 selectionArgs 参数，可以实现根据 ID 查询单条数据。

第 42~57 行重写了 insert( ) 方法，其中第 49 行代码调用 SQLiteDatabase 对象的 insert( ) 方法进行数据添加，并返回 ID；该方法要求返回新增数据的 URI，因此，在第 50 行调用 Uri. parse( ) 方法将以 ID 结尾的内容 URI 解析成 Uri 对象。

update( ) 和 delete( ) 方法代码与上述方法基本相同，此处不再赘述。

(2) 在【案例 2-6】项目的 AndroidManifest.xml 中对 MyProvider 进行声明。

在 <application></application> 标签中，加入如下代码。

```
1  <provider
2      android:authorities = "com.example.sqlitedemo2"
3      android:name = "com.example.sqlitedemo2.MyProvider"
4      android:exported = "true"/>
```

在上述代码中，使用 <provider> 标签对 MyProvider 进行声明，其中，"android：authorities" 属性值为 MyProvider 的权限，"android：name" 属性值为 MyProvider 类的全名，"android：exported" 属性值为 true 则代表 MyProvider 是可以被其他应用程序访问的。

(3) 创建新的项目 "UserContentResolverDemo"，该项目将访问案例【2-6】项目中共享的数据。

(4) 在新建项目的 activity_main.xml 布局文件中，编写布局代码。

在布局中加入 4 个按钮，分别用于添加一条数据、查询全部数据、更新一条数据、删除一条数据。布局代码较为简单，此处不再列出。

(5) 在 MainActivity.java 中，编写功能逻辑代码。

```
1  public class MainActivity extends AppCompatActivity implements
2                                  View.OnClickListener{
3      private Button bt_insert,bt_query,bt_update,bt_delete;
4      private ContentResolver resolver ;
5      private String newId;
6      @Override
7      protected void onCreate(Bundle savedInstanceState) {
8          super.onCreate(savedInstanceState);
9          setContentView(R.layout.activity_main);
10         initView();
11         resolver = getContentResolver();
12     }
13
14     public void initView(){
15         //省略控件初始化代码
16     }
17     @Override
18     public void onClick(View v) {
19         switch (v.getId()){
20             case R.id.bt_insert:{
21                 //添加数据
22                 Uri uri = Uri.parse("content://com.example.sqlitedemo2/users");
23                 ContentValues values = new ContentValues();
24                 values.put("username","systemAdmin");
25                 values.put("password","12356789");
26                 values.put("age",30);
```

```java
27             Uri newRri = resolver.insert(uri,values);
28             newId = newRri.getPathSegments().get(1);
29             Log.e("获取用户共享数据",newId);
30             break;
31         }
32         case R.id.bt_query:{
33             //查询数据
34             Uri uri = Uri.parse("content://com.example.sqlitedemo2/users");
35             Cursor cursor = resolver.query(uri,null,null,null,null);
36             if(cursor ! = null){
37                 while(cursor.moveToNext()){
38                     String username = cursor.getString(
39                             cursor.getColumnIndex("username"));
40                     String password = cursor.getString(
41                             cursor.getColumnIndex("password"));
42                     int age = cursor.getInt(cursor.getColumnIndex("age"));
43                     Log.d("获取用户共享数据",
44                             username+","+password+","+age);
45                 }
46             }
47             cursor.close();
48             break;
49         }
50         case R.id.bt_update:{
51             //更新数据
52             Uri uri = Uri.parse("content://com.example.sqlitedemo2/users/"
53                     +newId);
54             ContentValues values = new ContentValues();
55             values.put("username","systemAdmin");
56             values.put("password","rootroot");
57             values.put("age",30);
58             int updateRows = resolver.update(uri,values,null,null);
59             Log.e("获取用户共享数据",updateRows+"");
60             break;
61         }
62         case R.id.bt_delete:{
63             //删除数据
64             Uri uri = Uri.parse("content://com.example.sqlitedemo2/users/"
65                     +newId);
66             int deleteRows = resolver.delete(uri,null,null);
67             Log.e("获取用户共享数据",deleteRows+"");
68             break;
69         }
70     }
71 }
72 }
```

在上述代码中，第 18~77 行的 onClick( ) 方法中实现了 4 个按钮的单击事件。

第 20~32 行是添加一条数据时需要进行的操作，首先调用 Uri. parse( ) 方法将内容 URI 解析为 Uri 对象，然后将要添加的数据放入 ContentValues 对象中，调用 ContentResolver 的 insert( ) 方法执行添加操作；insert( ) 方法返回了一个 Uri 对象，其中包含了新增记录的 ID，第 30 行将 ID 取出，并通过 Log. e( ) 方法在日志中打印出来。

第 33~52 行是查询全部数据时需要进行的操作，首先，同样将内容 URI 解析成 Uri 对象，然后调用 query( ) 方法查询数据，并返回 Cursor 对象；接下来遍历 Cursor 对象，取出数据，并通过 Log. d( ) 方法打印出来。

第 53~65 行是更新一条数据时需要进行的操作，这里重点关注第 52 行，在内容 URI 的后面增加了一个 newId，这个 ID 就是单击"添加"按钮时返回的新增数据 ID，这个 Uri 代表需要更新的是新增的这条数据。

第 66~75 行是删除一条数据时需要进行的操作，具体代码与前面讲解的代码基本一致。

在运行程序时，需要先运行【案例 2-6】中的项目，然后运行 UserContentResolverDemo 项目。单击"保存"按钮，打印日志如图 2-35 所示。日志中打印了新增数据的 ID。

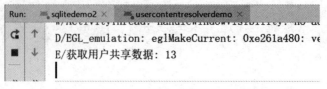

图 2-35　单击"保存"按钮日志打印情况

继续单击"查询"按钮，打印日志如图 2-36 所示。日志中列出了所有用户信息。

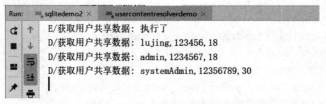

图 2-36　单击"查询"按钮日志打印情况

单击"更新"按钮并查询全部数据，打印日志如图 2-37 所示。日志中列出了所有用户信息，从打印信息可以看到，用户 systemAdmin 的密码更新为"rootroot"。

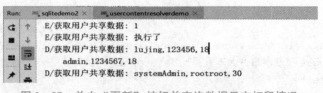

图 2-37　单击"更新"按钮并查询数据日志打印情况

最后，单击"删除"按钮并查询全部数据，打印日志如图 2-38 所示。日志中列出了所有用户信息，从打印信息可以看到，新增的用户信息已经删除。

图2-38 单击"删除"按钮并查询数据日志打印情况

任务实施

1. 任务分析

本任务需对题目信息进行数据库存储,并从数据库中读取题目信息,实现答题功能。根据任务需求,可以设计题目表用于存储题目信息,具体表结构见表2-6。

表2-6 题目表结构

序号	字段名	类型	是否允许为空	主键	说明
1	id	integer	否	是	题目ID,自动编号
2	questionstem	text	否		题干
3	optionA	text	否		选项A
4	optionB	text	否		选项B
5	optionC	text			选项C
6	optionD	text			选项D
7	correctoptions	text	否		正确选项,用0、1、2、3表示,如正确选项为A、B,该字段值为"0,1"
8	type	integer	否		题目类型,1为单选,2为多选

根据表结构创建数据表,插入题目信息,并随机读取5条数据,转换为List集合。

2. 实现步骤

在任务2的基础上,继续完成本任务功能。

(1)创建数据库。

创建名为"db"的包,并在其中创建DataBaseHelper类,并在onCreate()方法中创建数据表"Questions",具体代码如【代码2-11】所示。

【代码2-11】 DataBaseHelper.java

```
1  public class MyDataBaseHelper extends SQLiteOpenHelper {
2      //建表语句
3      public static final String CREATE_QUESTIONS = "create table
4  Questions ("
5              + "id integer primary key autoincrement,"
6              + "questionstem text,"
7              + "optionA text,"
```

```
8               + "optionB text,"
9               + "optionC text,"
10              + "optionD text,"
11              + "correctoptions text,"
12              + "type integer)";
13      private Context mContext;
14
15      public MyDataBaseHelper(Context context,String name,
16      SQLiteDatabase.CursorFactory factory,int version){
17          super(context,name,factory,version);
18          mContext = context;
19      }
20
21      @Override
22      public void onCreate(SQLiteDatabase db){
23          db.execSQL(CREATE_QUESTIONS);
24          Toast.makeText(mContext,"数据库创建成功!",
25                  Toast.LENGTH_SHORT).show();
26      }
27      @Override
28      public void onUpgrade(SQLiteDatabase db,int oldVersion,int newVersion){
29
30      }
31  }
```

（2）创建数据库操作类。

在 db 包中，创建数据库操作类 QuestionsDao.java，实现插入数据、查询数据、将查询结果 Cursor 转换为 List 集合的相关方法，具体代码如【代码 2-12】所示。

【代码 2-12】 QuestionsDao.java

```
1   public class QuestionsDao{
2       private MyDataBaseHelper helper;
3       public QuestionsDao(Context context){
4           helper = new MyDataBaseHelper(context,"LearnToPass.db",null,1);
5       }
6       //添加题目
7       public long insertQuestion(Question question){
8           SQLiteDatabase db = helper.getWritableDatabase();
9           ContentValues values = new ContentValues();
10          String[] options = question.getOptions();
11          values.put("questionstem",question.getQuestionStem());
12          values.put("optionA",options[0]);
13          values.put("optionB",options[1]);
14          values.put("optionC",options[2]);
15          values.put("optionD",options[3]);
```

```
16          values.put("correctoptions",question.getCorrectOptions());
17          values.put("type",question.getQuestionTypes());
18          long id = db.insert("questions",null,values);
19          db.close();
20          return id;
21      }
22      //随机查询5道题目
23      public ArrayList queryRandFive(){
24          ArrayList list = new ArrayList();
25          SQLiteDatabase db = helper.getWritableDatabase();
26          //随机查询5道题目SQL语句
27          String sql = "select * from questions order by random() limit 5";
28          Cursor cursor = db.rawQuery(sql,null);    //调用rawQuery()进行查询
29          list = convertFromCursor(cursor);
30          return list;
31      }
32      //取出查询记录返回题目集合
33      private ArrayList convertFromCursor(Cursor cursor){
34          ArrayList list = new ArrayList();
35          if (cursor ! = null && cursor.moveToNext()){
36              //通过游标遍历整个查询结果集
37              do {
38                  Question question = new Question();
39                  question.setQuestionStem(cursor.getString(1));
40                  String[] options = new String[]{
41                      cursor.getString(cursor.getColumnIndex("optionA")),
42                      cursor.getString(cursor.getColumnIndex("optionB")),
43                      cursor.getString(cursor.getColumnIndex("optionC")),
44                      cursor.getString(cursor.getColumnIndex("optionD"))};
45                  question.setOptions(options);
46                  question.setCorrectOptions(cursor.getString(
47                      cursor.getColumnIndex("correctoptions")));
48                  question.setQuestionTypes(cursor.getInt(
49                          cursor.getColumnIndex("type")));
50                  list.add(question);
51              } while (cursor.moveToNext());
52          }
53          return list;
54      }
55  }
```

(3) 改写答题界面QuestionActivity中的数据获取代码。

在QuestionActivity.java的initQuestions()方法中,当第一次调用该方法时需要创建题目对象,调用QuestionsDao类的insertQuestion()方法将题目插入数据库;再次答题时,则仅需调用QuestionsDao类的queryRandFive()方法查询题目即可。

为了区分是否是第一次调用，可以使用 SharedPreferences 存储布尔型的 isFirst，如果 isFirst默认为"true"，第一次调用结束后，isFirst 为"false"，具体代码如下。

【代码2-13】 QuestionActivity. java

```java
1   public class QuestionActivity2 extends AppCompatActivity{
2
3       //省略原变量定义
4       //新增 QuestionsDao 引用变量定义
5       private QuestionsDao dao;
6       @Override
7       protected void onCreate(Bundle savedInstanceState) {
8           super.onCreate(savedInstanceState);
9           setContentView(R.layout.activity_answer1);
10          dao = new QuestionsDao(this);
11          //初始化题目
12          initQuestions();
13          //初始化控件
14          initView();
15          //省略加载题目代码
16      }
17      //省略 initView()方法代码
18      //省略 replaceFragment()方法代码
19
20      private void initQuestions() {
21          //读取 isFirst 的值
22          SharedPreferences spf = getPreferences(MODE_PRIVATE);
23          boolean isFirst = spf.getBoolean("isFirst",true);
24          if(isFirst) {
25              //isFrist 为 true 代表第一次答题
26              SharedPreferences.Editor editor = spf.edit();
27              editor.putBoolean("isFirst",false);
28              editor.commit();
29              //第1题
30              Question question1 = new Question();
31              question1.setQuestionStem("第十四届全国冬季运动会第一次" +
32                      "在冬运会上设置_____季竞技项目。");
33              question1.setOptions(new String[]{"春","夏","秋","冬"});
34              question1.setCorrectOptions("1");
35              question1.setQuestionTypes(1);
36              dao.insertQuestion(question1);
37              //省略其余添加题目代码
38          }
39          //随机读取5道题目并以集合形式返回
40          questions = dao.queryRandFive();
41      }
42      //省略显示答题结束代码
43  }
```

任务反思

编写并运行程序，将在代码编写及程序调试过程中出现的异常信息、产生原因及解决方法记录在下方。

问题 1：_____

产生原因：_____

解决方法：_____

问题 2：_____

产生原因：_____

解决方法：_____

任务总结及巩固

小白：又到了学习每门编程语言必备的数据库部分了，相比之前我们用过的数据库，SQLite 数据库的应用可以说很简单了。

师兄：因为 SQLite 数据库是应用在移动设备上的，这就决定了它的小巧和便捷。

小白：SQLiteDatabase 提供的增、删、改、查方法让数据库操作方便了很多，即便不会写 SQL 语句，也能进行数据库操作。

师兄：这话可不对！SQL 语句是我们进行软件开发的基本功，一定要熟练掌握。在 SQLite 数据库这部分，数据库升级是我们要格外注意的。一定不要轻易使用删除数据表、删除数据库的语句。数据库中保存的是很重要的应用数据，在数据时代，数据就是生产力。数据的泄露和删除有可能给用户带来巨大的损失。我们不是经常能在技术公众号中看到"程序员删库跑路"的案例吗？这真是非常没有职业道德的做法。作为开发人员，我们一定要谨慎进行数据删除操作，同时，一定要保证用户数据安全，谨防数据泄露。

小白：Android 系统本身也通过 ContentProvider 保障了数据共享安全性。看来安全真是软件开发中需要重视的，我也得拉紧"数据安全"这根弦。

一、基础巩固

1. 对 SQLite 数据库进行添加操作时，可以使用 SQLiteDatabase 类中的（　　）方法。
   A. insert( )　　　B. query( )　　　C. update( )　　　D. execSQL( )

2. 下列方法中，（　　）方法可以返回 Cursor 对象。
A. execSQL( )　　　　　　　　　B. insert( )
C. update( )　　　　　　　　　　D. rawQuery( )
3. Android 中（　　）可以实现跨程序数据共享。
A. SQLiteDatabase　　　　　　　B. ContentProvider
C. ContentResolver　　　　　　　D. SharedPreferences
4. 下列关于 SQLite 数据库描述错误的是（　　）。
A. SQLiteOpenHelper 类一般用来创建和更新数据库
B. 当创建 SQLiteOpenHelper 对象时就创建数据库了
C. SQLiteOpenHelper 类中的 onCreate( )方法仅会在第一次安装程序时调用
D. 调用 getWritableDatabase( )方法可以获得 SQLiteDatabase 对象
5. 下列关于 ContentProvider 说法错误的是（　　）。
A. ContentProvider 不需要在 Manifest.xml 文件中声明
B. 如果想把 A 程序中的数据共享给其他应用，需要在 A 程序中创建 ContentProvider
C. 如果 B 程序想要访问 A 程序中共享的数据，需要借助 ContentResolver
D. 在 ContentProvider 中通过内容 URI 对数据进行操作

二、技术实践

1. 利用本任务中学习的数据库相关内容，参考图 2－39 所示的界面效果，完成一个题目管理程序，实现题目的增、删、改、查功能。

图 2－39　题目管理界面效果

2. 在题目管理程序中创建 ContentProvider 共享题目数据，尝试在"学习通关"项目中读取题目管理程序共享的题目数据。

任务 4 倒计时功能的实现

任务描述

实现答题界面左上角的倒计时功能,答题时间共 2 分钟,以"分:秒"的形式显示,如图 2-40 所示。

图 2-40 倒计时界面效果

任务学习目标

通过本任务需达到以下目标:
- 理解异步消息处理机制。
- 能够使用异步消息处理实现多线程编程。

技术储备

在 Java 程序开发中,当需要执行一些耗时的操作比如网络请求、图片下载时,会使用多线程进行处理,将操作放入子线程中运行。在 Android 开发中,也可以使用多线程。

在下面场景中,需要生成 100 个随机数,在页面上滚动显示。如果在主线程中,也就是 MainActivity 中实现此功能,将导致主线程阻塞,程序无法正常运行。因此,可以在子线程中实现此功能,具体代码如下。

```
1   public class MainActivity extends AppCompatActivity {
2
3       private TextView textview;
4       @Override
5       protected void onCreate(Bundle savedInstanceState) {
6           super.onCreate(savedInstanceState);
7           setContentView(R.layout.activity_main);
8           textview = findViewById(R.id.textview);
9           new Thread(new Runnable() {
10              @Override
11              public void run() {
12                  for(int i = 0; i < 100; i++) {
13                      double radom = Math.random() * 1000;
14                      textview.setText(Double.toString(radom));
15                      try {
16                          Thread.sleep(1000);
17                      } catch (InterruptedException e) {
18                          e.printStackTrace();
19                      }
20                  }
21              }}).start();
22      }
23  }
```

在代码中,第 9~21 行创建了一个子线程,在匿名内部类中实现了 Runnable 接口中的 run() 方法,在方法中循环生成随机数并更新到 Textview 控件上。运行程序,程序崩溃,提示图 2-41 所示异常信息。

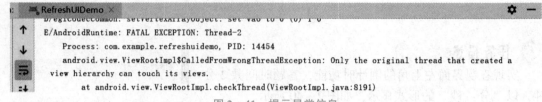

图 2-41 提示异常信息

这个异常是由于在子线程中更新 UI（界面）导致的。Android 的 UI 是线程不安全的，要更新应用程序中的 UI 元素，必须在主线程中进行。那么，像前面的场景中必须在子线程中执行耗时的任务，并根据任务的执行结果更新相应的 UI 控件，应该如何实现呢？

为了解决前面遇到的问题，Android 中提供了一套异步消息处理机制，解决了在子线程中更新 UI 的问题。

一、认识异步消息处理

所谓异步消息就是发送一个消息，不需要等待返回，随时可以再发送下一个消息。

异步消息处理线程启动后会进入一个无限的循环体之中，每循环一次，从其内部的消息队列中取出一个消息，然后回调相应的消息处理函数，执行完成一个消息后则继续循环，若消息队列为空，线程则会阻塞等待。

二、Android 异步消息处理机制

Android 中的异步消息处理主要由 4 个部分组成：Message、Handler、MessageQueue 和 Looper。下面对这 4 个部分进行简要介绍。

（一）Message

Message 是在线程之间传递的消息，它可以在内部携带少量信息，用于在不同线程之间交换数据。Message 的 what、arg1、arg2 字段可以用来携带一些整型数据，obj 字段可以携带 Object 对象。

（二）Handler

Handler 是消息的处理者，也是线程之间的通信兵，它主要用于发送和处理消息。发送消息一般使用 Handler 的 sendMessage() 方法，发出的消息经过一系列的处理，最终会传递到 Handler 的 handleMessage() 方法。

（三）MessageQueue

MessageQueue 是消息队列，它主要用于存放所有由 Handler 发送过来的消息，这部分消息会一直在消息队列中等待被处理。每个线程中只会有一个 MessageQueue 对象。

（四）Looper

Looper 是每个线程中 MessageQueue 的管家，调用 Looper 的 loop() 方法后，就会进入一个无限循环当中；每当发现 MessageQueue 中存在一条消息，就会将其取出，并传递到 handleMessage() 方法当中。每个线程中也只会有一个 Looper 对象。

图 2-42 描述了异步消息处理机制的整个流程。首先，需要在主线程中创建一个 Handler 对象，并重写 handleMessage( )方法。当子线程中需要创建 UI 操作时，就创建一个 Message 对象，并通过 Handler 将消息发送出去。这条消息会被添加到 MessageQueue 队列中等待被处理。Looper 会一直尝试从 MessageQueue 中取出待处理的消息，并调用 Handler 对象的 dispatchMessage( )方法分发回 handleMessage( )方法中进行处理。

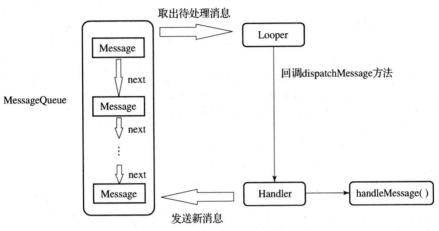

图 2-42　异步消息处理机制示意图

利用异步消息处理机制，我们可以对前面的生成随机数场景的代码进行改写。生成 20 个随机数，并在主界面控件上滚动显示的具体代码如下。

```
1  public class MainActivity extends AppCompatActivity {
2      private TextView textview;
3      private int i = 0;
4      private Handler handler = new Handler(){
5          @Override
6          public void handleMessage(Message msg) {
7              switch (msg.what) {
8                  case 0:
9                      textview.setText("已产生 20 个随机数结束");
10                     break;
11                 case 1:
12                     int radom = msg.arg1;
13                     textview.setText(Integer.toString(radom));
14                     break;
15                 default:
16             }
17         }
18     };
19     @Override
20     protected void onCreate(Bundle savedInstanceState) {
21         super.onCreate(savedInstanceState);
```

```
22          setContentView(R.layout.activity_main);
23          textview = findViewById(R.id.textview);
24          new Thread(new Runnable() {
25              @Override
26              public void run() {
27                  Message message = new Message();
28                  i++;
29                  if(i<=20) {
30                      int radom = (int)(Math.random()*1000);
31                      message.what = 1;
32                      message.arg1 = radom;
33                      handler.sendMessage(message);
34                      handler.postDelayed(this,1000);
35                  }else{
36                      message.what = 0;
37                      handler.sendMessage(message);
38                  }
39              }
40          }).start();
41      }
42  }
```

在上述代码中，第24~40行创建并启动了子线程，在线程的run()方法中，创建Message对象，如果计数变量i小于等于20，生成随机数，将Message对象的what字段设为1，生成的随机数赋值给arg1字段，并通过handler对象的sendMessage()方法发送消息到消息队列，最后调用handler对象的postDelayed()方法，该方法有两个参数，第1个参数为Runnable对象，第2个参数设为1 000毫秒，实现每隔1秒执行一次run()方法；如果i大于20，则将Message对象的what字段设为0，并发送消息。

第4~18行定义了Handler对象并重写handleMessage()方法进行消息处理，通过判断Message对象的what字段决定如何操作。如果what字段为0，则意味着已生成20个随机数，显示"已产生20个随机数结束"；如果what字段为1，则继续获取arg1字段中的随机数，显示在控件上。

任务实施

1. 任务分析

本任务需要采用异步消息处理机制实现倒计时功能。

在子线程中实现时间以秒为单位的递减，将当前时间存入Message对象的字段中，通过Handler的sendMessage()方法将消息发到消息队列。消息队列中的消息被分发到Handler的handleMessage()，handMessage()对消息进行处理。

2. 实现步骤

在任务3完成的QuestionActivity.java基础上，继续完成本任务功能。

（1）创建Handler对象并重写handleMessage()方法，具体代码如【代码2-14】所示。

【代码 2-14】 QuestionActivity.java

```
1   public class QuestionActivity extends AppCompatActivity{
2
3       //省略原变量定义
4       //新增本任务相关变量
5       private TextView spare_time;
6       private int time = 2 * 60;
7       private Handler handler = new Handler(){
8           @Override
9           public void handleMessage(@NonNull Message msg){
10              switch (msg.what){
11                  case 0:
12                      //倒计时结束
13                      showDialog();
14                      break;
15                  case 1:
16                      //从消息中取出当前分及秒,在控件上更新
17                      int minute = msg.arg1;
18                      int second = msg.arg2;
19                      spare_time.setText(minute + ":" + second);
20                      break;
21                  default:
22              }
23          }
24      };
25      //省略其他功能代码
26  }
```

(2) 创建并启动子线程,在线程中实现时间逐秒递减。

```
1   public class QuestionActivity extends AppCompatActivity{
2       //省略原变量定义
3       private TextView spare_time;
4       private int time = 2 * 60;     //共 2 分钟,120 秒
5       private class ThreadRunnable implements Runnable {
6           @Override
7           public void run(){
8               time--;     //每次执行 time 减 1
9               if(time == 0){
10                  //倒计时结束
11                  Message message = new Message();
12                  message.what = 0;   //0 代表倒计时结束
13                  handler.sendMessage(message);//发送消息
14              } else {
15                  //倒计时正在进行
16                  Message message = new Message();
17                  message.what = 1;
18                  message.arg1 = time/60;     //分钟
```

```
19              message.arg2 = time% 60;    //秒
20              handler.sendMessage(message);   //发送消息
21              handler.postDelayed(this,1000);//间隔1秒调用 Runnable 对象
22          }
23      }
24  }
25  //省略其他功能代码
26  }
```

📀 任务反思

编写并运行程序，将在代码编写及程序调试过程中出现的异常信息、产生原因及解决方法记录在下方。

问题1：

产生原因：

解决方法：

问题2：

产生原因：

解决方法：

📀 任务总结及巩固

师兄：小白，异步消息处理机制能够实现 Android 的多线程，你能总结一下异步消息处理机制吗？

小白：我觉得异步消息处理机制的重点是要弄清楚4个角色加处理流程。Message 是消息，能够携带数据；Handler 发送并处理消息；MessageQueue 是消息的存放处；Looper 像是永动机，不断拿出消息分发给 Handler 进行处理。在子线程中进行耗时处理及消息发送，回到主线程进行界面更新。

师兄：总结得不错。过程你理解了，但是到了实际应用中，要根据不同的功能需求，具体问题具体分析、实现，这才是难点，而这也是需要在实践中多看、多想、多练的。

在这个任务中，我们使用了异步消息处理机制完成了倒计时功能，这实际上就是一个定时器。Timer 也可以完成这个任务，这个就留给你回去找找资料啦！

一、基础巩固

1. 下列说法正确的是（　　）。
   A. 网络请求、下载图片等耗时的操作可以放在主线程中进行
   B. Android 中允许在子线程中进行 UI 更新
   C. 发送一个消息不需要等待返回是同步消息处理
   D. Android 中采用异步消息处理机制实现多线程
2. Android 的异步消息处理中，处理消息由（　　）负责。
   A. Message　　　　B. Handler　　　　C. MessageQueue　　D. Looper
3. 下列方法中，（　　）用来发送消息。
   A. sendMessage( )　　　　　　　B. handleMessage( )
   C. postDelayed( )　　　　　　　D. post( )
4. Android 的异步消息处理中，取出消息并传递给 handlMessage( ) 方法由（　　）负责。
   A. Message　　　　　　　　　　B. Handler
   C. MessageQueue　　　　　　　　D. Looper
5. Android 的异步消息处理中，存放 Handler 发送过来的消息由（　　）负责。
   A. Message　　　　　　　　　　B. Handler
   C. MessageQueue　　　　　　　　D. Looper

二、技术实践

使用本任务学习的内容，实现一个从 10 开始的倒数功能。

学习目标达成度评价

了解对模块的自我学习情况，完成表 2-7。

表 2-7　学生目标达成度评价表

序号	学习目标	学生自评
1	能够运用 ListView、SimpleAdapter、BaseAdapter 实现页面列表功能	□能够熟练实现列表 □通过查看以往代码及课本，能够基本实现列表 □不理解列表的实现方法
2	能够运用 RadioButton 及 CheckBox 等控件实现单选、复选功能	□能够正确理解并熟练使用控件 □通过查看以往代码及课本，能够基本实现单选、复选功能 □不会两种控件的使用

续表

序号	学习目标	学生自评
3	能够正确理解 Fragment 的作用并创建 Fragment	□能够正确理解并熟练创建 Fragment □通过查看以往代码及课本，能够基本实现 Fragment □不会 Fragment 的创建
4	能够通过 SQLite 数据库实现数据的存储	□能够熟练运用 SQLite 数据库实现数据存储 □通过查看以往代码及课本，能够基本实现数据存储 □不知道如何存储数据
5	能够理解 ContentProvider 的使用并运用它实现数据共享	□能够熟练运用 ContentProvider 实现数据共享 □通过查看以往代码及课本，能够基本实现数据共享 □不知道如何使用 ContentProvider
6	理解异步消息处理机制，能够使用 Handler 实现简单的异步消息处理	□能够熟练进行异步消息处理 □通过查看以往代码及课本，能够基本实现异步消息处理 □不知道如何实现异步消息处理
评价得分			
学生自评得分（20%）	学习成果得分（60%）	学习过程得分（20%）	模块综合得分

（1）学生自评得分。

每个学习目标有 3 个选项，选项第 1 个选项得 100 分，选项第 2 个选项得 70 分，选项第 3 个选项得 50 分，学生自评得分为各项学习目标得分的平均分。

（2）学习成果得分。

教师根据学生阶段性测试及模块学习成果的完成情况酌情赋分，满分为 100 分。

（3）学习过程得分。

教师根据学生的其他学习过程表现，如到课情况、参与课程讨论等情况酌情赋分，满分为 100 分。

功能模块 3

学习模块的实现

说在前面

小白：接下来我们要进一步开发哪些功能呢？

师兄：我们目前完成的功能中数据都是从数据库中读取的，而实际应用中的数据大多来源于服务器，因此，应用程序需要能够实现网络通信功能；同时，在大部分应用中，页面最上方会有一个标题栏提供搜索、回到上一个页面等功能，在接下来的模块中我们将重点研究这两方面的内容。

小白：现在很多 App 打开后都会看到上方和下方的导航栏，这个又是怎么实现的呢？

师兄：这也是我们在这一部分需要实现的，我们将利用这些知识完成学习模块的设计与实现，具体分为以下 3 个任务。

任务 1：标题栏及导航栏的设计与实现。

任务 2：顶部标签栏及内容列表的设计与实现。

任务 3：通过网络获取资讯数据。

功能需求描述

完成如图 3-1~图 3-3 所示的学习模块。

（1）页面顶部实现包含搜索框、总积分的顶部工具栏，底部包含导航栏，单击不同导航切换到相应功能页面。

（2）页面中间以列表方式呈现学习内容，左右滑动页面可以在时政要闻、科技发展、技术技能等栏目间切换。

（3）页面通过网络通信从服务器获取，并对获取到的数据进行解析，显示在页面上。

图3-1 学习模块—时政要闻

图3-2 学习模块—科技发展

图3-3 电台模块

学习目标

1. 知识目标

（1）掌握顶部工具栏及底部导航栏的实现方法。

（2）掌握 ViewPager、RecyclerView、TabLayout 控件的使用方法。

（3）掌握通过 HttpURLConnection 访问网络的方法。

（4）掌握 JSON 数据解析方法。

2. 技能目标

（1）能够根据界面需要，利用 ToolBar 等控件实现顶部工具栏和底部导航栏。

（2）能够利用 ViewPager、RecycleView、TabLayout 控件实现顶部标签栏、列表等功能效果。

（3）能够访问网络获取 JSON 数据并进行数据解析。

任务1　标题栏及导航栏的设计与实现

任务描述

根据学习模块的参考界面，在学习界面中添加如图3-4所示的顶部标题栏以及如图3-5所示的底部导航栏，单击不同的导航标签可以跳转至不同页面或不同 Fragment。

图3-4 学习界面顶部工具栏

图3-5 学习界面底部导航栏

任务学习目标

通过本任务需达到以下目标：

➢ 能够通过 Toolbar 实现顶部标题栏功能。
➢ 能够使用 BottomNavigationView 实现底部导航栏功能。

技术储备

一、使用 Toolbar 实现顶部标题栏

顶部标题栏是页面中的常见部分，在顶部标题栏中通常会包含比如返回上一页、扫一扫、搜索、菜单等功能。

Toolbar 是 Google 公司在 Android 5.0 版本中推出的导航控件，可以方便地创建顶部标题栏。Toolbar 实际上是一个 ViewGroup，可以包含任何的子控件，用户可以根据自己的产品需求设计不同的标题栏。

（一）Toolbar 的基本使用

图 3-6 为一个 ToolBar 的基本组成，包括左侧导航图标、标题、子标题等。具体的创建步骤如下。

Toolbar 的
基本使用

图 3-6 Toolbar 的基本组成

1. 隐藏页面自带标题栏

创建页面的默认样式中包含 ActionBar，ActionBar 是 Android 5.0 以前的顶部标题栏，由于其使用不够灵活，现已被 Toolbar 取代。要创建 Toolbar 需要将 ActionBar 隐藏。下面介绍两种隐藏 ActionBar 的方法，分别通过主题设置及 Java 代码来实现。

（1）将应用主题设为 "NoActionBar"。

在 AndroidManifest.xml 里面设置应用的 android：theme 属性值为 "@style/Theme.AppCompat.Light.NoActionBar"。采用这种方式可以将应用中所有页面的 ActionBar 去掉。

```
1  < application
2          android:allowBackup = "true"
3          android:icon = "@mipmap/ic_launcher"
4          android:label = "@string/app_name"
5          android:supportsRtl = "true"
6          android:theme = "@style/Theme.AppCompat.Light.NoActionBar" >
7          ……
8  </application >
```

（2）在 Java 代码中调用方法，隐藏 ActionBar。

在 Activity 的 onCreate（）方法中，调用 getSupportActionBar（）方法获取 ActionBar，并继续调用 hide（）方法将 ActionBar 隐藏。该方法的调用语句需放在 setContentView（）方法调用之后。

```
1  @Override
2      protected void onCreate(Bundle savedInstanceState) {
3          super.onCreate(savedInstanceState);
4          setContentView(R.layout.activity_study);
5          getSupportActionBar().hide();
6          ……
7  }
```

2. 在页面布局文件中加入 Toolbar

与前面学习的 Button、ListView 等控件不同，在加入 Toolbar 时，需要指明 Toolbar 所在的包，即"androidx.appcompat.widget.Toolbar"。可以根据功能需求在 Toolbar 标签中添加其他控件，形成个性化的标题栏。

```
1  <?xml version = "1.0" encoding = "utf-8"?>
2  <LinearLayout xmlns:android = "http://schemas.android.com/apk/res/android"
3      android:layout_width = "match_parent"
4      android:layout_height = "match_parent"
5      android:orientation = "vertical">
6      <androidx.appcompat.widget.Toolbar
7          android:id = "@+id/toolbar"
8          android:layout_width = "match_parent"
9          android:layout_height = "wrap_content">
10         <!-- 其他控件 -->
11     </androidx.appcompat.widget.Toolbar>
12 </LinearLayout>
```

3. 在 Activity 代码中设置 Toolbar

Toolbar 提供了多个属性及相应的设置方法用于指定标题栏的风格，具体设置代码如下。

```
1  public class MainActivity extends AppCompactActivity {
2      private Toolbar toolBar;
3      @Override
4      protected void onCreate(Bundle savedInstanceState) {
5          super.onCreate(savedInstanceState);
6          setContentView(R.layout.activity_main);
7          toolBar = findViewById(R.id.toolbar);
8          //设置 Logo 图标
9          toolBar.setLogo(R.drawable.logo);
10         //设置左侧导航图标
11         toolBar.setNavigationIcon(R.drawable.ic_back);
12         //设置标题
13         toolBar.setTitle("标题栏");
14         //设置标题文字颜色
15         toolBar.setTitleTextColor(Color.BLACK);
16         //设置子标题
17         toolBar.setSubtitle("子标题栏");
```

```
18          //设置背景颜色
19          toolBar.setBackgroundResource(R.color.colorAccent);
20          //设置 ToolBar 对象
21          setSupportActionBar(toolBar);
22          toolbar.setNavigationOnClickListener(
23                      new OnClickListener(){
24              @Override
25              public void onClick(View v) {
26                  finish();
27              }}
28          );
29      }
30  }
```

上述代码中,通过调用 Toolbar 对象的 "set×××( )" 方法进行标题栏各部分的设置,第 20~26 行添加了左侧导航图标单击事件监听,注意:这部分代码需要写在 setSupportAction-Bar( )方法后才能起作用。

(二) 为 Toolbar 添加菜单

在有些 Toolbar 上会看到图 3-7 中右侧的菜单按钮,这需要通过添加菜单来实现。对于 Toolbar 上没有空间完全显示的菜单项或设置为不显示的菜单项,则隐藏在右侧的溢出菜单中,单击图标则弹出菜单,具体实现步骤如下。

为 Toolbar 添加菜单

图 3-7 Toolbar 中的菜单

1. 添加菜单资源

在 res 目录下,创建 "menu" 资源文件夹。在文件夹中,添加 "menu_main.xml" 文件。

```
1  <menu xmlns:android = http://schemas.android.com/apk/res/android
2      xmlns:app = "http://schemas.android.com/apk/res-auto">
3      <item android:id = "@+id/action_join"
4          android:title = "加入"
5          android:icon = "@drawable/ic_join"
6          app:showAsAction = "always"/>
7      <item android:id = "@+id/action_notifications"
8          android:title = "提示"
9          android:icon = "@drawable/ic_notifications"
10         app:showAsAction = "ifRoom"/>
11     <item android:id = "@+id/action_quit"
12         android:title = "退出"
13         android:orderInCategory = "100"
14         android:icon = "@drawable/ic_quit"
15         app:showAsAction = "never"/>
16 </menu>
```

上述菜单定义中，包含 3 个菜单项，分别为"加入""提示""退出"，"app：showAsAction"属性设置了菜单项在标题栏上的展示位置，其属性值有以下 5 种。

- always：总是在标题栏上显示菜单图标。
- ifRoom：如果右侧有空间，该项就显示在标题栏上。
- never：从不在标题栏上直接显示，一直放在溢出菜单列表中。
- withText：如果能在标题栏上显示，除了显示图标，还要显示文字说明。
- collapseActionView：操作视图要折叠为一个按钮，单击该按钮再展开操作视图，主要用于设置 SearchView。

2. 加载菜单

在 Activity 中通过回调方法 onCreateOptionsMenu（Menu menu）加载菜单，具体代码如下。

```
1  @Override
2  public Boolean onCreateOptionsMenu(Menu menu) {
3      getMenuInflater().inflate(R.menu.menu_main,menu);
4      return true;
5  }
```

3. 为菜单项添加单击事件

在 Activity 中通过为 Toolbar 对象添加 onMenuItemClickListener 监听器处理菜单项单击事件，具体代码如下。

```
1  toolBar.setOnMenuItemClickListener(new Toolbar.OnMenuItemClickListener()
2  {
3      @Override
4      public boolean onMenuItemClick(MenuItem item) {
5          switch(item.getItemId()) {
6              case R.id.action_join:
7                  Toast.makeText(MainActivity.this,"单击加入",
8                                  Toast.LENGTH_SHORT).show();
9                  break;
10             case R.id.action_notifications:
11                 Toast.makeText(MainActivity.this,"单击提示",
12                                 Toast.LENGTH_SHORT).show();
13                 break;
14             case R.id.action_quit:
15                 Toast.makeText(MainActivity.this,"单击退出",
16                                 Toast.LENGTH_SHORT).show();
17                 break;
18         }
19         return true;
20     }
21 });
```

在上述方法中，菜单项单击事件由 onMenuItemClick() 方法处理，其参数为当前单击的菜单项对象，通过 MenuItem 对象的 getItemId() 获取菜单项 id 与菜单中定义的 id 进行匹配，确定哪个菜单项被单击。注意：以上代码需要放在 setSupportActionBar() 之后，否则不起作用。

（三）在 Toolbar 中添加搜索框（SearchView）

在常见的应用中，在标题栏的中间通常会有一个搜索框。随着移动应用功能的不断丰富，搜索框中还可以进行二维码扫描等功能。

SearchView 与 ListView 等控件类似，是 Android 中提供的搜索框控件，它能够提供一个用户界面，用于搜索查询。通过 SearchView 可以在 Toolbar 中添加基础搜索框，具体步骤如下。

为 Toolbar 添加搜索框

1. 在菜单资源文件中加入搜索项

在菜单文件中，通过"app：actionViewClass"属性将 SearchView 作为一个菜单项添加到 ToolBar 中。

```
1  <menu xmlns:android = http://schemas.android.com/apk/res/android
2      xmlns:app = "http://schemas.android.com/apk/res-auto">
3      <item android:id = "@+id/action_search"
4          android:title = "搜索"
5          android:icon = "@drawable/ic_search"
6          app:showAsAction = "always"
7          app:actionViewClass = "androidx.appcompat.widget.SearchView"/>
8      ……
9  </menu>
```

2. 设置 SearchView 的样式

在 Activity 的 onCreateOptionsMenu(Menu menu) 方法中获取并设置 SearchView。

```
1  @Override
2      public boolean onCreateOptionsMenu(Menu menu) {
3          getMenuInflater().inflate(R.menu.menu_main, menu);
4          //获取搜索框菜单项
5          MenuItem item = menu.findItem(R.id.action_search);
6          //获取 SearchView 对象
7          SearchView searchView = (SearchView)item.getActionView();
8          //设置搜索栏的默认提示
9          searchView.setQueryHint("请输入商品名称");
10         //默认搜索栏为展开状态
11         searchView.setIconified(false);
12         //设置搜索框事件监听器
13         searchView.setOnQueryTextListener(new SearchView.OnQueryTextListener() {
14             //输入搜索内容后提交时触发
15             @Override
16             public boolean onQueryTextSubmit(String query) {
17                 Toast.makeText(MainActivity.this,"您输入的文本为"+query,
```

```
18                                          Toast.LENGTH_SHORT).show();
19                  return false;
20              }
21              //输入内容改变时触发
22              @Override
23              public boolean onQueryTextChange(String newText) {
24                  return false;
25              }
26          });
27          return true;
28      }
29
```

在上述代码中,第 4~7 行获取 SearchView 对象;第 8~11 行设置搜索框的提示信息及是否默认为展开状态;第 13~28 行为 SearchView 对象添加事件监听,其中,onQueryText-Submit(String query)是当文本提交时的回调,onQueryTextChange(String newText)是当输入文本发生改变时的回调。

按照上面学习的内容及编写的代码,我们就可以实现如图 3-8 所示的顶部标题栏,单击放大镜图标展开搜索框,即产生如图 3-9 所示的效果。

图 3-8 顶部标题栏运行效果

图 3-9 搜索框展开效果

二、使用 BottomNavigationView 实现底部导航栏

创建底部导航栏

底部导航功能是移动应用中的常见功能,图 3-10 和图 3-11 就是支付宝和淘宝 App 的底部导航效果。移动设备的屏幕空间有限,通过导航功能,尤其是导航与 Fragment 的结合能够在一个 Activity 中集成多个功能模块,使移动应用的功能更加丰富。

图 3-10 支付宝 App 底部导航

图 3-11 淘宝 App 底部导航

底部导航栏有多种实现方法,使用前面学习的基本控件与布局,比如 TextView、LinearLayout、RadioButton 等就可以实现底部导航栏。我们也可以使用 Google 公司推出的 BottomNavigationView 控件实现底部导航栏。

下面我们通过 BottomNavigationView 控件实现图 3-5 中学习模块的底部导航效果。

(一)导入依赖包

BottomNavigationView 控件是 Android 官方为开发者提供的一种 Material 组件,使用 Material 组件可以满足开发者日常开发 UI 的需求,提高开发效率。在使用 Material 组件之前,需要在项目中导入依赖包。在项目的 build.gradle 文件中,dependencies 中引入依赖"com.google.android.material:material:1.2.0"。

```
1  dependencies {
2      implementation fileTree(dir:'libs',include: ['*.jar'])
3      implementation'androidx.appcompat:appcompat:1.0.2'
4      implementation'androidx.constraintlayout:constraintlayout:1.1.3'
5      testImplementation'junit:junit:4.12'
6      androidTestImplementation'androidx.test.ext:junit:1.1.0'
7      androidTestImplementation'androidx.test.espresso:espresso-core:3.1.1'
8      implementation'com.google.android.material:material:1.2.0'
9  }
```

(二)创建底部导航菜单

底部导航栏中要显示的图标和图标下面的文字是通过 menu 菜单设置的。在 res/menu 目录中创建菜单文件"nav_menu.xml"。将底部导航栏中的 5 个导航设置为 5 个菜单项,具体代码如下。

```
1  <?xml version="1.0" encoding="utf-8"?>
2  <menu xmlns:android="http://schemas.android.com/apk/res/android"
3      xmlns:app="http://schemas.android.com/apk/res-auto">
4      <item
5          android:id="@+id/points"
6          android:icon="@drawable/points"
7          android:title="积分" />
8      <item
9          android:id="@+id/answer"
10         android:icon="@drawable/answer"
11         android:title="答题" />
12     <item
13         android:id="@+id/study"
14         android:icon="@drawable/study"
15         android:title="学习" />
16     <item
17         android:id="@+id/radio"
18         android:icon="@drawable/radio"
19         android:title="电台" />
20     <item
21         android:id="@+id/tv"
22         android:icon="@drawable/tv"
23         android:title="电视台" />
24 </menu>
```

上述代码中,"android:icon"属性为底部导航栏所显示的图标,"android:title"属性为图标下面的文字。

(三) 将 BottomNavigationView 控件加入布局

在界面的布局文件中加入 BottomNavigationView 控件,让控件位于页面的底部。

```xml
1   <?xml version = "1.0" encoding = "utf-8"?>
2   <RelativeLayout xmlns:android = "http://schemas.android.com/apk/res/android"
3       xmlns:app = "http://schemas.android.com/apk/res-auto"
4       xmlns:tools = "http://schemas.android.com/tools"
5       android:layout_width = "match_parent"
6       android:layout_height = "match_parent"
7       tools:context = ".activity.StudyActivity" >
8
9       <com.google.android.material.bottomnavigation.BottomNavigationView
10          android:id = "@+id/bottom_nav"
11          android:layout_width = "match_parent"
12          android:layout_height = "50dp"
13          android:layout_alignParentBottom = "true"
14          android:background = "#ffffff"
15          app:itemBackground = "@null"
16          app:itemIconTint = "@drawable/selector_nav"
17          app:itemTextColor = "@drawable/selector_nav"
18          app:labelVisibilityMode = "labeled"
19          app:menu = "@menu/nav_menu" />
20  </RelativeLayout>
```

在上述代码中,第 10~20 行向布局中加入了 BottomNavigationView 控件,注意加入时需要指定 BottomNavigationView 的所在包;第 13 行将 BottomNavigationView 设置在页面的底端。

第 14 行将 BottomNavigationView 的背景设为白色;第 15 行用于设置导航栏中菜单项的背景,属性值为 "@null" 表示去掉背景。

第 16 行 app:itemIconTint 属性用于设置导航栏中图标的颜色;第 17 行 app:itemTextColor 属性用于设置导航栏中文字的颜色,这两个属性值可以设置为 selector 选择器,以实现选择和未选择菜单项时图标及文字颜色的变化。

第 18 行 app:labelVisibilityMode = "labeled" 可以解决底部导航菜单项大于 3 个时文字不显示的问题;第 19 行 app:menu 属性为控件设置菜单项。

(四) 添加 BottomNavigationView 单击效果

经过上面 3 个步骤,运行程序就可以在界面底端加入底部导航栏了,但此时单击底部导航还没有具体的功能效果。通常情况下,单击底部导航可以跳转到新的页面,也可以出现不同的功能内容,但在新的页面中底部导航栏也会一直存在。第 1 种情况下,可以直接通过 Intent 完成页面跳转;第 2 种情况下,需要将 BottomNavigationView 与 Fragment 配合使用。下面我们学习第 2 种情况的实现方法。

1. 创建导航对应的 Fragment

根据功能要求,为每个导航所对应的功能创建 Fragment。这里我们都用 Fragment1、Fragment2、Fragment3、Fragment4、Fragment5 来替代。

2. 在布局文件中加入 FrameLayout

在布局文件中 BottomNavigationView 控件的上方加入一个 FrameLayout 布局，用于添加 Fragment。

```
1   <?xml version = "1.0" encoding = "utf-8"?>
2   <RelativeLayout xmlns:android = "http://schemas.android.com/apk/res/android"
3       xmlns:app = "http://schemas.android.com/apk/res-auto"
4       xmlns:tools = "http://schemas.android.com/tools"
5       android:layout_width = "match_parent"
6       android:layout_height = "match_parent"
7       tools:context = ".activity.StudyActivity">
8       <FrameLayout
9           android:id = "@+id/main_con"
10          android:layout_width = "match_parent"
11          android:layout_height = "match_parent"
12          android:layout_marginTop = "50dp"
13          android:layout_marginBottom = "50dp" />
14      <!--省略 BottomNavigationView 控件-->
15  </RelativeLayout>
```

3. 在 Activity 中实现单击导航菜单项切换 Fragment

创建 initBottomNavigation() 方法用于设置底部导航，并在 onCreate() 方法中调用该方法。在该方法中需要完成以下几项操作。

（1）创建 Fragment 对象，并将其添加到 FrameLayout 中。

（2）添加导航菜单项单击事件，根据单击菜单项，显示对应的 Fragment。

```
1   public classMainActivity extends AppCompatActivity {
2   
3       private List<Fragment> fragmentList = new ArrayList<>();
4       privateBottomNavigationView bottomNavigationView;
5   
6       @Override
7       protected void onCreate(Bundle savedInstanceState) {
8           super.onCreate(savedInstanceState);
9           setContentView(R.layout.activity_study);
10          //初始化底部导航
11          initBottomNavigation();
12      }
13  
14      private void initBottomNavigation() {
15          /*
16           * 将所有的碎片放到集合当中
17           * 获取 FragmentManager 通过 getSupportFragmentManager 获得
18           * 添加 Fragment 并提交
19           */
```

```java
20        fragmentList.add(newFragment1());
21        fragmentList.add(newFragment2());
22        fragmentList.add(newFragment3());
23        fragmentList.add(newFragment4());
24        fragmentList.add(newFragment5());
25        final FragmentManager fragmentManager =
26                            getSupportFragmentManager();
27        fragmentManager.beginTransaction()
28                .add(R.id.main_con,fragmentList.get(0),"积分")
29                .add(R.id.main_con,fragmentList.get(1),"答题")
30                .add(R.id.main_con,fragmentList.get(2),"学习")
31                .add(R.id.main_con,fragmentList.get(3),"电台")
32                .add(R.id.main_con,fragmentList.get(4),"电视台")
33                .commit();
34        //进入主界面的时候显示"学习"界面,隐藏其他 Fragment
35        fragmentManager.beginTransaction()
36                .hide(fragmentList.get(0))
37                .hide(fragmentList.get(1))
38                .hide(fragmentList.get(3))
39                .hide(fragmentList.get(4))
40                .commit();
41        //设置底部导航选中"学习"菜单项
42        bottomNavigationView.setSelectedItemId(R.id.study);
43        //添加导航菜单项事件监听
44        bottomNavigationView.setOnNavigationItemSelectedListener(
45        new BottomNavigationView.OnNavigationItemSelectedListener() {
46            @Override
47            public boolean onNavigationItemSelected(MenuItem item) {
48                switch (item.getItemId()) {
49                    case R.id.points:
50                        fragmentManager.beginTransaction()
51                                .hide(fragmentList.get(1))
52                                .hide(fragmentList.get(2))
53                                .hide(fragmentList.get(3))
54                                .hide(fragmentList.get(4))
55                                .show(fragmentList.get(0))
56                                .commit();
57                        return true;
58                    //选择其余菜单项时的代码与上相似
59                }
60                return false;
61            }
62        });
63    }
64 }
```

在上述代码中，第 3 行创建了 Fragment 集合对象，第 20~24 行创建 Fragment 对象并添加到集合中。

第 28~33 行获取 fragmentManager 对象，调用 beginTransaction( ) 方法开启事务，将 5 个 Fragment 对象通过 add( ) 方法添加到布局中的 FrameLayout 布局中，并提交事务。

第 35~40 行开启事务，调用 hide( ) 方法将除了要作为默认显示的 Fragment 之外的其他 Fragment 隐藏起来。

第 42 行代码调用 bottomNavigationView 对象的 setSelectedItemId( ) 方法设置默认选中的导航菜单项。

第 43~62 行为 bottomNavigationView 对象添加菜单项选中事件监听器，重写 onNavigationItemSelected( ) 方法，通过判断选中菜单项 id，调用 Fragment 对象的 show( ) 方法显示对应的 Fragment，并将其他 Fragment 隐藏。

任务实施

1. 任务分析

本任务中"学习"界面需要实现顶部标题栏及底部导航栏。

顶部标题栏可以通过 ToolBar 来实现。可以通过在前面 ToolBar 实现代码中重新设置左侧图标、添加 SearchView、将"我的"设置为菜单项来基本实现界面效果，但在细节上会有所不同。也可以通过在布局中自行添加所需控件来实现，采用这种方法也能够实现更为丰富的顶部标题栏效果。

底部导航栏的实现方法在前面已经基本完成，仅需要将"积分"及"答题"导航菜单项被单击后的效果修改为跳转至功能模块 2 中实现的页面即可。

2. 实现步骤

在功能模块 2 完成的项目基础上，继续进行本任务的实现。

（1）在 activity 包中，创建新的 Activity，Activity 名称为 "StudyActivity"，Layout 名称为 "activity_study"。

（2）在布局文件 activity_study.xml 中，编写布局代码。

在布局代码中，需要在顶部加入 ToolBar，其中，通过线性布局加入 1 个 SearchView 和 2 个 TextView；下方加入 FrameLayout 和 BottomNavigationView。

【代码 3 – 1】 activity_study.xml

```
1   <?xml version = "1.0" encoding = "utf - 8"?>
2   <RelativeLayout xmlns:android = "http://schemas.android.com/apk/res/android"
3       xmlns:app = "http://schemas.android.com/apk/res - auto"
4       xmlns:tools = "http://schemas.android.com/tools"
5       android:layout_width = "match_parent"
6       android:layout_height = "match_parent"
7       tools:context = ".activity.StudyActivity" >
8       <androidx.appcompat.widget.Toolbar
9           android:id = "@ + id/main_toolbar"
10          android:layout_width = "match_parent"
```

```xml
11          android:layout_height = "50dp"
12          android:background = "@color/red"
13          app:contentInsetLeft = "0dp"
14          app:contentInsetStart = "0dp" >
15          <LinearLayout
16              android:layout_width = "match_parent"
17              android:layout_height = "match_parent"
18              >
19              <androidx.appcompat.widget.SearchView
20                  android:id = "@+id/searchView"
21                  android:layout_width = "0dp"
22                  android:layout_weight = "4"
23                  android:layout_height = "match_parent"
24                  android:background = "@drawable/shape_search"
25                  android:layout_margin = "5dp" />
26              <TextView
27                  android:id = "@+id/tb_points"
28                  android:layout_width = "0dp"
29                  android:layout_height = "match_parent"
30                  android:layout_weight = "1"
31                  android:gravity = "center"
32                  android:text = "4200 分"
33                  android:textColor = "@color/white"
34                  android:textSize = "16sp"
35                  android:textStyle = "bold" />
36              <TextView
37                  android:id = "@+id/tb_me"
38                  android:layout_width = "0dp"
39                  android:layout_weight = "1"
40                  android:layout_height = "match_parent"
41                  android:gravity = "center"
42                  android:text = "我的"
43                  android:textColor = "@color/white"
44                  android:textSize = "16sp"
45                  android:textStyle = "bold" />
46          </LinearLayout>
47      </androidx.appcompat.widget.Toolbar>
48      <!—省略 FrameLayout 及 BottomNavigationView 代码 -->
49  </RelativeLayout>
```

上述代码中，在 Toolbar 标签中，加入一个水平线性布局，其中，包含 1 个 SearchView 控件、2 个 TextView 控件。SearchView 的背景设置为一个圆角且填充为白色的 shape。加入 FrameLayout 及 BottomNavigationView 的代码在前面已经实现了，此处将其省略。

3. 实现底部导航功能

本任务底部导航栏所需菜单及主要逻辑功能代码在前面的学习中已经基本实现。仅需将前面的代码放入 StudyActivity. java 中,并在细节上进行修改。

(1) 在 fragment 包中,创建 StudyFragment、AudioFragment 及 VideoFragment,并替换前面代码中的 Fragment3、Fragment4 和 Fragment5。

(2) 单击"积分"及"答题"导航菜单项,将跳转至功能模块 2 中实现的功能页面,需将这两个菜单项的单击事件代码进行修改。

【代码 3-2】 StudyActivity. java

```
1   public class StudyActivity extends AppCompatActivity {
2       private Toolbar toolbar;
3       private List < Fragment > fragmentList = new ArrayList < >();
4       private BottomNavigationView bottomNavigationView;
5       @Override
6       protected void onCreate(Bundle savedInstanceState) {
7           super.onCreate(savedInstanceState);
8           setContentView(R.layout.activity_study);
9           //初始化视图
10          initView();
11          //初始化底部导航
12          initBottomNavigation();
13      }
14      private void initView() {
15          //获得对应的控件
16          bottomNavigationView = findViewById(R.id.bottom_nav);
17          toolbar = findViewById(R.id.main_toolbar);
18          //设置控件的样式
19          toolbar.setLogo(R.drawable.logo2);
20      }
21      private void initBottomNavigation() {
22          fragmentList.add(newStudyFragment());
23          fragmentList.add(newAudioFragment());
24          fragmentList.add(newVideoFragment());
25          final FragmentManager fragmentManager =
26                              getSupportFragmentManager();
27          fragmentManager.beginTransaction()
28                  .add(R.id.main_con,fragmentList.get(0),"学习")
29                  .add(R.id.main_con,fragmentList.get(1),"电台")
30                  .add(R.id.main_con,fragmentList.get(2),"电视台")
31                  .commit();
32          //进入主界面的时候显示"学习"界面
33          fragmentManager.beginTransaction()
34                  .hide(fragmentList.get(1))
35                  .hide(fragmentList.get(2))
36                  .commit();
37          //设置底部导航选中"学习"按钮
38          bottomNavigationView.setSelectedItemId(R.id.study);
39          bottomNavigationView.setOnNavigationItemSelectedListener(
```

```
40            new BottomNavigationView.OnNavigationItemSelectedListener() {
41                @Override
42                public boolean onNavigationItemSelected(MenuItem item) {
43                    switch (item.getItemId()) {
44                        case R.id.points:
45                            Intent intent_points = new Intent(
46                                    StudyActivity.this,PointsActivity.class);
47                            startActivity(intent_points);
48                            return true;
49                        case R.id.answer:
50                            Intent intent_answer = new Intent(
51                                    StudyActivity.this,QuestionActivity.class);
52                            startActivity(intent_answer);
53                            return true;
54                        case R.id.study:
55                            fragmentManager.beginTransaction()
56                                    .hide(fragmentList.get(1))
57                                    .hide(fragmentList.get(2))
58                                    .show(fragmentList.get(0))
59                                    .commit();
60                            return true;
61                        case R.id.radio:
62                            fragmentManager.beginTransaction()
63                                    .hide(fragmentList.get(0))
64                                    .hide(fragmentList.get(2))
65                                    .show(fragmentList.get(1))
66                                    .commit();
67                            return true;
68                        case R.id.tv:
69                            fragmentManager.beginTransaction()
70                                    .hide(fragmentList.get(0))
71                                    .hide(fragmentList.get(1))
72                                    .show(fragmentList.get(2))
73                                    .commit();
74                            return true;
75                    }
76                    return false;
77                }
78            });
79       }
80  }
```

任务反思

编写并运行程序，将在代码编写及程序调试过程中出现的异常信息、产生原因及解决方法记录在下方。

问题 1：_____

产生原因：_____

_____

解决方法：_____

_____

问题 2：_____

产生原因：_____

_____

解决方法：_____

任务总结及巩固

小白：顶部标题栏和底部导航栏是现在移动应用中不可缺少的部分。

师兄：随着用户对应用功能及操作性要求越来越高，在有限的页面空间中需要展现的功能也越来越多。

小白：仅仅一个 ToolBar 就将"回退"按钮、菜单、搜索控件集为一体，一个小小的空间就能实现多种功能。而通过底部导航栏将多个功能部分在一个页面上显示出来，让原本有限的界面空间显示了更多内容。

师兄：没错，你可以用来实现界面功能的工具又多了两个。除了 ToolBar 和 BottomNavigationView，顶部标题栏和底部导航功能都还有其他的实现方法。你可以上网查查资料，看看还有哪些方法可以实现这两种功能，它们又各有哪些优缺点？

一、基础巩固

1. 在菜单中，（　　）属性可以设置菜单项始终显示在标题栏中。

   A. app：showAsAction = "always"　　　B. app：showAsAction = "never"

   C. app：showAsAction = "ifroom"　　　D. app：showAsAction = "withText"

2. 下列方法中，（　　）用来为 ToolBar 中的菜单项添加单击事件。

   A. setOnMenuItemClickListener( )　　　B. setOnItemClickListener( )

   C. setOnClickListener( )　　　D. setOnMenuItemSelectedListener( )

3. 下列关于 Toolbar 说法错误的是（　　）。

   A. 在菜单文件中，通过 app：actionViewClass 属性添加搜索框

   B. 在 Toolbar 中不显示的菜单项可以在右侧溢出菜单中显示

C. onCreateOptionsMenu（Menu menu）方法可以加载菜单

D. ToolBar 无需在布局中添加

4. 下列关于 BottomNavigationView 说法错误的是（　　）。

A. setSelectedItemId( )方法可以设置底部导航选中的菜单项

B. BottomNavigationView 通常可以与 Fragment 配合使用

C. BottomNavigationView 中导航图片和文字不是通过添加菜单实现的

D. 使用 BottomNavigationView 需要在 build.gradle 中添加依赖

二、技术实践

选择一个你经常使用的应用，利用本任务中学习的内容实现应用的顶部标题栏和底部导航功能。

任务 2　顶部标签栏及内容列表的设计与实现

任务描述

完成学习导航对应的功能界面，如图 3-1、图 3-2 所示。

界面中顶部为标签栏，包括"时政要闻""科技发展""技术技能"等标签。每个标签对应下方的内容列表。左右滑动屏幕可以在 3 个标签及对应内容间切换，如图 3-12 所示。

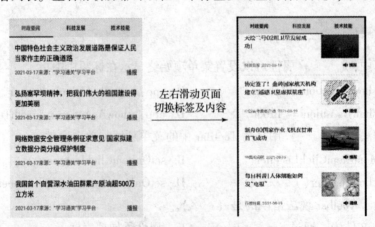

图 3-12　滑动切换效果

任务学习目标

通过本任务需达到以下目标：
➢ 能够使用 ViewPager 控件实现翻页效果。
➢ 能够使用 TabLayout 控件实现顶部标签栏。
➢ 能够使用 RecyclerView 控件实现内容列表功能。

技术储备

一、使用 ViewPager 控件实现翻页效果

使用 ViewPager 实现滑动翻页

ViewPager 是一种可以让用户左右切换当前视图的控件。在应用程序中，页面切换、轮播图、启动引导页等功能都可以通过 ViewPager 实现。

ViewPager 类直接继承自 ViewGroup 类，是一个容器类，可以在其中添加其他的 View 类，也可以添加 Fragment 和 Activity。

与 ListView 类似，ViewPager 类需要借助 PagerAdapter 适配器为它提供要显示的内容 View。ViewPager 经常和 Fragment 一起使用，并且提供了 PagerAdapter 适配器的子类 FragmentPagerAdapter，用于将 Fragment 放入 ViewPager。

下面我们就来实现一个具有 3 个 Fragment 的滑动翻页效果。

（一）在主界面布局文件中添加 ViewPager 控件

与其他控件的用法相同，ViewPager 在使用时也需要先加入界面的布局文件中，并且需要指明具体的包 "androidx.viewpager.widget.ViewPager"，具体代码如下。

```
1  <?xml version="1.0" encoding="utf-8"?>
2  <LinearLayout xmlns:android="http://schemas.android.com/apk/res/android"
3      xmlns:app="http://schemas.android.com/apk/res-auto"
4      xmlns:tools="http://schemas.android.com/tools"
5      android:layout_width="match_parent"
6      android:layout_height="match_parent"
7      android:orientation="vertical"
8      tools:context=".fragment.PassFragment">
9      <!--翻页视图-->
10     <androidx.viewpager.widget.ViewPager
11         android:id="@+id/viewpager"
12         android:layout_width="match_parent"
13         android:layout_height="match_parent" />
14 </LinearLayout>
```

（二）创建适配器类继承 FragmentPagerAdapter

FragmentPagerAdapter 作为 PagerAdapter 的子类已经实现了几个必需的方法，我们只需要实现 getCount() 方法及 getItem() 方法。同时，在适配器类中创建构造方法，传入 FrameManager 对象及放入 ViewPager 的 Fragment 对象集合。

```java
1  public class MyPagerAdapter extends FragmentPagerAdapter {
2
3      private List<Fragment> fragmentList;
4      public MyPagerAdapter(FragmentManager fm,List<Fragment> fragmentList) {
5          super(fm);
6          this.fragmentList = fragmentList;
7      }
8      //获取当前滑动到的 Fragment 对象
9      @Override
10     public Fragment getItem(int position) {
11         return fragmentList.get(position);
12     }
13     //获取 Fragment 的个数
14     @Override
15     public int getCount() {
16         return fragmentList.size();
17     }
18 }
```

上述代码中，getItem( )方法返回当前滑动到的 Fragment 对象，参数为当前 Fragment 的下标，可以通过参数 position 从 Fragment 集合中获得当前 Fragment 对象返回的下标；getCount( )方法返回 Fragment 的个数。

（三）为 ViewPager 对象设置适配器

在 MainActivity.java 中，获取 ViewPager 对象；创建 ViewPager 中的 Fragment 对象，并添加到集合中；创建 MyPagerAdapter 适配器对象，并通过 ViewPager 对象的 setAdapter( )方法进行设置。

```java
1  public class MainActivity extends AppCompatActivity {
2      private ViewPager viewpager;
3      @Override
4      protected void onCreate(Bundle savedInstanceState) {
5          super.onCreate(savedInstanceState);
6          setContentView(R.layout.activity_main);
7          //1. 获取 ViewPager 控件
8          viewpager = findViewById(R.id.viewpager);
9          //2. 准备 Fragment 集合
10         List<Fragment> fragmentList = new ArrayList<Fragment>();
11         FirstFragment firstFragment = new FirstFragment();
12         SecondFragment secondFragment = new SecondFragment();
13         ThirdFragment thirdFragment = new ThirdFragment();
14         fragmentList.add(firstFragment);
15         fragmentList.add(secondFragment);
16         fragmentList.add(thirdFragment);
17
```

```
18        //3. 创建 Adapter 对象
19        MyPagerAdapter adapter =
20          new MyPagerAdapter(getSupportFragmentManager(),fragmentList);
21        //4. 设置 Adapter
22        viewpager.setAdapter(adapter);
23      }
24    }
```

(四) 对 ViewPager 设置事件监听

我们可以通过 addOnPageChangeListener( )方法为 ViewPager 添加 OnPageChangeListener 事件监听,当页面滑动时触发事件。

```
1   viewpager.addOnPageChangeListener(new ViewPager.OnPageChangeListener() {
2       @Override
3       public void onPageScrolled(int position,float positionOffset,
4                                   int positionOffsetPixels) {
5         //页面滑动时调用本方法
6       }
7       @Override
8       public void onPageSelected(int position) {
9         //当滑动到某个页面时调用本方法,position 为滑动到的页面位置
10        Toast.makeText(MainActivity.this,"这是第" + position +1 +
11              "个 Fragment",Toast.LENGTH_LONG).show();
12      }
13      @Override
14      public void onPageScrollStateChanged(int state) {
15        //当页面滚动状态发生变化时调用本方法
16      }
17  });
```

在处理 ViewPager 页面改变事件时,需要实现 3 个方法。

(1) onPageScrolled( )方法在页面滑动时调用,在滑动停止之前,会一直调用此方法,第 1 个参数为当前页面,即单击滑动的页面,第 2 个参数是当前页面偏移的百分比,第 3 个参数是当前页面偏移的像素位置。

(2) onPageSelected( )方法在页面滑动结束后调用,参数 position 是当前选中的页面位置。

(3) onPageScrollStateChanged( )方法在页面滑动状态改变时调用,参数 state 有 3 个值:SCROLL_STATE_DRAGGING 表示用户手指按在屏幕上并且开始拖动的状态、SCROLL_STATE_IDLE 表示滑动动画做完的状态、SCROLL_STATE_SETTLING 表示手指离开屏幕的状态。

通过上述 3 个方法的实现,可以在滑动页面时产生需要的效果。

经过上面 4 个步骤,我们就完成了一个可以滑动展示 3 个 Fragment 的页面效果,具体运行效果如图 3 - 13 所示。页面首先显示第一个 Fragment,向左滑动屏幕,显示第二个 Fragment,滑动结束调用 onPageSelected( )方法,显示 Toast 提示信息。

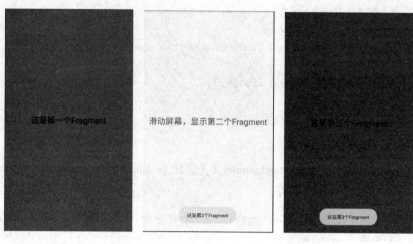

图 3-13 滑动屏幕切换 Fragment 并显示提示信息

长知识：

FragmentPagerAdapter 与 FragmentStatePagerAdapter 的区别

除了 FragmentPagerAdapter 可以将 Fragment 放入 ViewPager 外，FragmentStatePagerAdapter 也有同样的作用，它也是 PagerAdpter 的子类。使用 FragmentPagerAdapter 时，每一个 Fragment 都将保存在内存中，适合数量较少的页面；而 FragmentStatePagerAdapter 只保留当前页面，当离开页面后，页面就会被消除，释放资源，在页面需要时，生成新的页面，适合大量页面。

二、使用 TabLayout 控件实现顶部标签栏

TabLayout 是 com. google. android. material 包中一种实现顶部导航的常用控件。TabLayout 提供了一个水平布局来展示标签，一般结合 ViewPager 和 Fragment 实现滑动的标签选择器。

下面我们为上一部分实现的滑动翻页效果添加顶部标签，使每一个 Fragment 对应一个标签，滑动屏幕的同时，标签页跟随发生变化。

（一）在 build. gradle 中加入依赖

TabLayout 与 BottomNavigationView 都是 Material 组件，使用时，需要在项目的 build. gradle 中注入依赖 "com. google. android. material：material：1.2.0"。具体代码在 BottomNavigationView 的学习中已经完成，此处不再赘述。

注意：Android 开发技术不断迭代发展，新功能层出不穷，各种组件版本不断更新，在注入依赖时，需要注意包名后的版本号是否为最新版本号。同时，Android 支持库也经历了从 V4、V7、V13 到现在的 androidx，我们现在用到的很多控件、Fragment、AppCompatActivity 都出自 androidx 包中。

（二）在布局文件中加入 TabLayout 控件

在上一部分完成的主界面布局中，在 ViewPager 控件的前面加入 TabLayout 控件。

```xml
1  <?xml version="1.0" encoding="utf-8"?>
2  <LinearLayout xmlns:android="http://schemas.android.com/apk/res/android"
3      xmlns:app="http://schemas.android.com/apk/res-auto"
4      xmlns:tools="http://schemas.android.com/tools"
5      android:layout_width="match_parent"
6      android:layout_height="match_parent"
7      android:orientation="vertical"
8      tools:context=".fragment.PassFragment">
9      <!--顶部标签-->
10     <com.google.android.material.tabs.TabLayout
11         android:id="@+id/tablayout"
12         android:layout_width="match_parent"
13         android:layout_height="50dp"
14         android:background="@color/light_yellow"
15         app:tabIndicatorColor="@color/red"
16         app:tabSelectedTextColor="@color/red"
17         app:tabTextColor="@color/black" />
18     <!--翻页视图-->
19     <androidx.viewpager.widget.ViewPager
20         android:id="@+id/viewpager"
21         android:layout_width="match_parent"
22         android:layout_height="match_parent" />
23 </LinearLayout>
```

在上述布局文件中，第 11~18 行加入了 TabLayout 控件，其中，定义了 3 个属性，app:tabSelectedTextColor 设置选中标签字体的颜色，app:tabIndicatorColor 设置标签文字下方的指示器颜色，app:tabTextColor 设置标签默认字体颜色。

（三）对 ViewPager 的适配器类进行改写

在实现 ViewPager 时，我们创建了 MyPagerAdapter 类。在与 TabLayout 配合使用时，需要通过构造方法传入每个 Fragment 对应的标签集合，并重写 getPageTitle()方法，返回当前 Fragment 对应的标签。注意：Fragment 集合与标签集合需要一一对应。

```java
1  public class MyPagerAdapter extends FragmentPagerAdapter {
2  
3      private List<Fragment> fragmentList;
4      private List<String> titles;
5      public MyPagerAdapter(FragmentManager fm,List<Fragment> fragmentList,
6      List<String> titles) {
7          super(fm);
8          this.fragmentList = fragmentList;
9          this.titles = titles;
```

```
10    }
11    //获取当前滑动到的Fragment对象
12    @Override
13    public Fragment getItem(int position){
14        return fragmentList.get(position);
15    }
16    //获取Fragment的个数
17    @Override
18    public int getCount(){
19        return fragmentList.size();
20    }
21    //获取当前Fragment对应的标签
22    @Override
23    public CharSequence getPageTitle(int position){
24        return titles.get(position);
25    }
26 }
```

（四）在Java代码中将TabLayout与ViewPager相关联

调用TabLayout对象的setupWithViewPager()方法，为TabLayout设置ViewPager对象。

```
1  public class MainActivity extends AppCompatActivity {
2  
3      private List<Fragment> fragments = new ArrayList<>();
4      private List<String> titles = new ArrayList<>();
5      private ViewPager viewPager;
6      private TabLayout tablayout;
7      @Override
8      protected void onCreate(Bundle savedInstanceState){
9          super.onCreate(savedInstanceState);
10         setContentView(R.layout.activity_main);
11         initView();
12     }
13     public void initView(){
14         //找到对应的控件
15         viewPager = findViewById(R.id.viewpager);
16         tablayout = findViewById(R.id.tablayout);
17         //省略原代码中ViewPager控件设置代码
18  
19         //关联TabLayout与ViewPager
20         tablayout.setupWithViewPager(viewPager);
21     }
22 }
```

运行程序，滑动屏幕会出现如图3-14所示的标签变化效果。

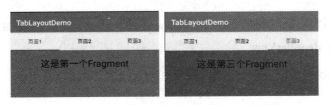

图 3-14 切换 Fragment 及标签栏变化效果

三、使用 RecyclerView 实现列表

RecyclerView 是在 Android 5.0 版本推出的一种功能强大的控件。RecyclerView 与 ListView 类似，也是以列表的形式展示数据，但是 RecyclerView 的功能非常强大，有更加丰富的展示效果，能够展示横向、竖向、网格、瀑布流等效果的列表；RecyclerView 的适配器的性能更加高效，具备强大的 Item 的复用机制，并能为 Item 添加动画效果。

RecyclerView 的使用

查文档找答案

查找 API 文档学习 RecyclerView 的常用方法，完成表 3-1。

表 3-1 RecyclerView 的常用方法

方法名称	功能描述
setAdapter( )	
setLayoutManager( )	
addItemDecoration( )	
removeItemDecoration( )	
setItemAnimator( )	
addOnItemTouchListener( )	
removeOnItemTouchListener( )	

下面我们将功能模块 2 中完成的积分列表改为用 RecyclerView 实现，学习 RecyclerView 的基本用法。其中，列表项布局是无论使用 RecyclerView 还是使用 ListView 都需要实现的，这里我们就直接使用前面完成的布局文件 "item_points.xml"。

（一）在 build.gradle 中加入依赖

目前 RecyclerView 所在的 support 被迁移到 androidx 中，所以需要在 app/build.gradle 的 dependencies 中引入依赖，具体代码如下。

```
implementation 'androidx.recyclerview:recyclerview:1.1.0'
```

(二)在布局文件中添加 RecyclerView 控件

在界面布局文件所需位置加入 RecyclerView 控件,注意,此处需要加上包名"androidx. recyclerview. widget. RecyclerView"。

```xml
1  <?xml version = "1.0" encoding = "utf-8"?>
2  <FrameLayout xmlns:android = "http://schemas.android.com/apk/res/android"
3      android:layout_width = "match_parent"
4      android:layout_height = "match_parent" >
5      <androidx.recyclerview.widget.RecyclerView
6          android:id = "@+id/recyclerview"
7          android:layout_width = "match_parent"
8          android:layout_height = "wrap_content"/>
9  </FrameLayout >
```

(三)创建自定义适配器继承 RecyclerView. Adapter

RecyclerView. Adapter 是 RecyclerView 的专用适配器。在创建自定义适配器时必须重写 RecyclerView. Adapter 的 getItemCount()、onCreateViewHolder() 和 onBindViewHolder() 方法,3 个方法的具体含义如下所示。

(1) getItemCount():获得列表项的数目。

(2) onCreateViewHolder():创建整个布局的视图持有者。

(3) onBindViewHolder():绑定每项的持有者。

```java
1  public class LinearAdapter extends RecyclerView.Adapter {
2      private List < PointsItem > dataList;   //数据列表
3      private LayoutInflater inflater;
4      private Context context;
5
6      public LinearAdapter(Context context,List < PointsItem > dataList) {
7          inflater = LayoutInflater.from(context);
8          this.dataList = dataList;
9          this.context = context;
10     }
11     @Override
12     public ViewHolder onCreateViewHolder(ViewGroup parent,int viewType) {
13         //加载列表项布局
14         View view = inflater.inflate(R.layout.item_points,parent,false);
15         ViewHolder holder = new ViewHolder(view);   //创建 ViewHolder 对象
16         return holder;
17     }
18     @Override
19     public void onBindViewHolder(RecyclerView.ViewHolder holder,
20                                   final int position) {
21         ViewHolder itemholder = (ViewHolder)holder;
```

```
22        PointsItem item = dataList.get(position);
23        //为列表项控件设置内容
24        itemholder.points_title.setText(item.getTitle());
25        itemholder.points_introduce.setText(item.getIntroduce());
26        //添加列表项上的按钮单击事件
27        itemholder.points_btn.setOnClickListener(
28          new View.OnClickListener() {
29            @Override
30            public void onClick(View v) {
31              switch(position){
32                case 3 :
33                  Toast.makeText(context,"每日答题的按钮被单击",
34                                 Toast.LENGTH_SHORT).show();
35                  break;
36              }
37            }
38        });
39    }
40    @Override
41    public int getItemCount() {
42        return dataList = = null ? 0 : dataList.size();
43    }
44    //创建 ViewHolder 类,获取控件对象
45    public class ViewHolder extends RecyclerView.ViewHolder {
46        private TextView points_title,points_introduce;
47        private Button points_btn;
48        ViewHolder(View view) {
49            super(view);
50            points_title = view.findViewById(R.id.points_title);
51            points_introduce = view.findViewById(R.id.points_introduce);
52            points_btn = view.findViewById(R.id.points_btn);
53        }
54    }
55 }
56
```

在上述代码中,第 48~57 行创建了 ViewHolder(视图持有者)类,创建并获取列表项控件对象。在列表项加载过程中,不再重复创建和获取控件对象,而是复用 ViewHolder 中的对象,提升了列表项加载的性能。

第 12~17 行重写了 onCreateViewHolder( )方法,加载列表项布局,创建并返回 ViewHolder对象。

第 19~39 行重写了 onBindViewHolder( )方法,实现列表项数据的设置、控件及列表项的事件处理。

(四) 在 Activity 中对 RecyclerView 进行设置

在 Activity 的 Java 文件中，获取 RecyclerView 对象，并对 RecyclerView 进行设置，具体代码如下。

RecyclerView 的设置

```
1  public class MainActivity extends AppCompatActivity {
2      publicRecyclerView recyclerView;
3      public List<PointsItem> list;
4      @Override
5      protected void onCreate(Bundle savedInstanceState) {
6          super.onCreate(savedInstanceState);
7          setContentView(R.layout.activity_main);
8          //准备数据
9          initData();
10         //找到对应的控件
11         recyclerView = findViewById(R.id.recyclerview);
12         //创建线性布局管理器
13         LinearLayoutManager manager = new LinearLayoutManager(this);
14         //设置线性布局的方向为垂直
15         manager.setOrientation(RecyclerView.VERTICAL);
16         //设置布局管理器
17         recyclerView.setLayoutManager(manager);
18         //创建适配器对象
19         LinearAdapter adapter = new LinearAdapter(this,list);
20         recyclerView.setAdapter(adapter);
21         //添加列表项分隔线
22         recyclerView.addItemDecoration(new DividerItemDecoration(this,
23             LinearLayoutManager.VERTICAL));
24     }
25     private void initData() {
26         //数据列表创建及积分项数据创建代码省略
27     }
28 }
```

在上述代码中，第 13~17 行创建了 LinearLayoutManager 对象，并通过 setLayoutManager() 方法设置到 RecyclerView 上。LayoutManager 是 RecyclerView 的精髓，它提供了 3 种布局管理器：线性布局管理器 (LinearLayoutManager)、网格布局管理器 (GridLayoutManager) 和瀑布流网格布局管理器 (StaggeredGridLayoutManager)。通过 setLayoutManager() 方法设置后，界面就会根据新布局刷新列表项。在上面的代码中使用了线性布局管理器，可以实现水平或垂直的列表形式。网格布局管理器及瀑布流网格布局管理器的实现效果如图 3-15 和图 3-16 所示。

第 19~20 行创建适配器对象并通过 setAdapter() 方法进行设置；第 22 行调用 addItemDecoration() 方法设置了列表项的分隔线效果，这里使用了 Android 官方提供的分隔线处理方案 DividerItemDecoration，我们也可以通过继承 RecyclerView.ItemDecoration 来实现自定义分隔线。

图 3-15 网格布局管理器实现效果　　　　图 3-16 瀑布流网格布局管理器实现效果

至此，我们就使用 RecyclerView 实现了积分列表。运行程序，可以呈现如图 3-17 所示的列表效果。

图 3-17 积分列表效果

任务实施

1. 任务分析

本任务需要实现在底部导航栏选择"学习"时显示如图 3-12 所示的带顶部标签栏的学习内容列表。由于每个底部导航菜单项对应的是一个 Fragment，本任务功能要求需要在"学习"菜单项对应的 Fragment，即 StudyFragment 中实现。

StudyFragment 中包含顶部标签栏和滑动翻页效果，需要 TabLayout 和 ViewPager 配合实现。页面中的列表显示在 ViewPager 中，这里我们将 RecyclerView 实现的列表放在 Fragment 中实现。

2. 实现步骤

在本模块任务 1 完成的项目基础上，继续进行本任务的实现。

（1）在 entity 包中创建新闻实体类 News。

在本部分列表中显示用于学习的新闻内容，提取属性标题（title）、内容（content）、日期（date）和来源（source）形成实体类。

【代码3-3】 News.java

```
1  public class News {
2      private String title;//题目
3      private String content;//内容
4      private String source;//来源
5      private String date;//日期
6  //省略构造方法及setter、getter方法 }
```

（2）在StudyFragment的布局文件fragment_study.xml中实现界面布局。

界面中包括一个TabLayout实现顶部标签栏，下方是一个ViewPager实现滑动翻页效果。

【代码3-4】 fragment_study.xml

```
1  <?xml version = "1.0" encoding = "utf - 8"?>
2  <LinearLayout xmlns:android = "http://schemas.android.com/apk/res/android"
3      xmlns:app = "http://schemas.android.com/apk/res - auto"
4      xmlns:tools = "http://schemas.android.com/tools"
5      android:layout_width = "match_parent"
6      android:layout_height = "match_parent"
7      android:orientation = "vertical"
8      tools:context = ".fragment.StudyFragment" >
9      <!-- 顶部标签 -->
10     <com.google.android.material.tabs.TabLayout
11         android:id = "@ + id/tab_study"
12         android:layout_width = "match_parent"
13         android:layout_height = "50dp"
14         android:background = "@color/light_yellow"
15         app:tabIndicatorColor = "@color/red"
16         app:tabSelectedTextColor = "@color/red"
17         app:tabTextColor = "@color/black" />
18     <!-- 翻页视图 -->
19     <androidx.viewpager.widget.ViewPager
20         android:id = "@ + id/vp_study"
21         android:layout_width = "match_parent"
22         android:layout_height = "wrap_content" />
23 </LinearLayout>
```

（3）创建StudySubFragment1、StudySubFragment2、StudySubFragment3用于显示时政内容、科技内容、技能内容列表。

ViewPager中放入的是多个Fragment，显示的时政内容等学习内容列表应该在Fragment中显示。本步骤中创建的多个Fragment用来显示学习内容列表，此处仅实现StudySubFragment1用来显示时政列表。

①在fragment_studysub1.xml中，编写StudySubFragment1的布局代码。

布局中仅包含一个RecyclerView控件。

【代码3-5】 fragment_studysub1.xml

```
1   <?xml version="1.0" encoding="utf-8"?>
2   <LinearLayout xmlns:android="http://schemas.android.com/apk/res/android"
3       xmlns:tools="http://schemas.android.com/tools"
4       android:layout_width="match_parent"
5       android:layout_height="match_parent"
6       tools:context=".fragment.PassSubpage1Fragment">
7       <!--RecyclerView-->
8       <androidx.recyclerview.widget.RecyclerView
9           android:id="@+id/rv_studysub1"
10          android:layout_width="match_parent"
11          android:layout_height="wrap_content" />
12  </LinearLayout>
```

②创建布局文件 fragment_study_item.xml，在其中编写时政列表项布局。

布局采用相对布局，其中包含 4 个 TextView 控件，用于存放标题、日期、来源及右侧的播报，具体代码可扫描二维码查看。

③在 adapter 包中，创建 RecyclerView 适配器 StudySubAdapter。StudySubAdapter 继承自 RecyclerView.Adapter，并重写了相关方法。

代码:fragment_study_item.xml

【代码3-6】 StudySubAdapter.java

```
1   public class StudySubAdapter extends
2                   RecyclerView.Adapter<StudySubAdapter.ViewHolder>{
3       private List<News> newsList;
4       private Context context;
5       public StudySubAdapter(Context context,List<News> newsList){
6           this.context=context;
7           this.newsList=newsList;
8       }
9       @Override
10      public ViewHolder onCreateViewHolder(ViewGroup parent,int viewType){
11          View view=LayoutInflater.from(context)
12                  .inflate(R.layout.fragment_study_item,parent,false);
13          ViewHolder holder=new ViewHolder(view);
14          return holder;
15      }
16      @Override
17      public void onBindViewHolder(ViewHolder holder,int position){
18          News news=newsList.get(position);
19          holder.title.setText(news.getTitle());
20          holder.date.setText(news.getDate());
21          holder.source.setText("来源:"+news.getSource());
22      }
23      @Override
24      public int getItemCount(){
```

```
25        return newsList.size();
26    }
27    class ViewHolder extends RecyclerView.ViewHolder {
28        TextView title;
29        TextView date;
30        TextView source;
31        public ViewHolder(View itemView) {
32            super(itemView);
33            title = itemView.findViewById(R.id.tv_study_title);
34            date = itemView.findViewById(R.id.tv_study_date);
35            source = itemView.findViewById(R.id.tv_study_source);
36        }
37    }
38 }
```

④在 StudySubFragment1.java 中,实现 RecyclerView 相关设置。

【代码 3-7】 StudySubFragment1.java

```
1  public class StudySubFragment1 extends Fragment {
2      private List <News> newsList = new ArrayList<>();
3      @Override
4      public void onCreate(Bundle savedInstanceState) {
5          super.onCreate(savedInstanceState);
6      }
7      @Override
8      public View onCreateView(LayoutInflater inflater,ViewGroup container,
9      Bundle savedInstanceState) {
10         View view = inflater.inflate(R.layout.fragment_studysub1,
11                                      container,false);
12         return view;
13     }
14     @Override
15     public void onViewCreated(View view,Bundle savedInstanceState) {
16         super.onViewCreated(view,savedInstanceState);
17         //初始化 RecyclerView
18         initRecyclerView(view);
19         //初始化数据
20         initData();
21     }
22     //初始化 RecyclerView
23     private void initRecyclerView(View view) {
24         RecyclerView recyclerView = view.findViewById(R.id.rv_studysub1);
25         //创建线性布局管理器
26         LinearLayoutManager manager =
27                             new LinearLayoutManager(getContext());
28         //设置布局管理器
```

```
29          recyclerView.setLayoutManager(manager);
30          StudySubAdapter adapter =
31                  new StudySubAdapter(getContext(),newsList);
32          recyclerView.setAdapter(adapter);
33      }
34      //初始化数据
35      private void initData(){
36          News news = new News();
37          news.setTitle("中国特色社会主义政治发展道路是保证人民当家作" + "主的正确道路");
38          news.setDate("2021 - 03 - 17");
39          news.setSource(""学习通关"学习平台");
40          newsList.add(news);
41          //省略其他数据初始化代码
42      }
43  }
44 }
```

（4）在 adapter 包中，创建 ViewPagerAdapter。

ViewPagerAdapter 继承自 FragmentPagerAdapter，由于 ViewPager 需要与 TabLayout 关联，需要重写 getPageTitle( )方法。

【代码 3 - 8】 ViewPagerAdapter.java

```
1  public class ViewPagerAdapter extends FragmentPagerAdapter {
2      private List <Fragment> fragments;    //viewpager 中 fragment 集合
3      private List <String> titles;          //TabLayou 中标签集合
4      public ViewPagerAdapter(FragmentManager fm,List <Fragment> fragments,
5      List <String> titles) {
6          super(fm);
7          this.fragments = fragments;
8          this.titles = titles;
9      }
10     @Override
11     public Fragment getItem(int position) {
12         return fragments.get(position);
13     }
14     @Override
15     public int getCount() {
16         return fragments.size();
17     }
18     @Override
19     public CharSequence getPageTitle(int position) {
20         return titles.get(position);
21     }
22 }
```

（5）在 StudyFragment.java 中，实现 TabLayout 与 ViewPager 的关联。

【代码 3 - 9】 StudyFragment.java

```java
public class StudyFragment extends Fragment {
    private List<Fragment> fragments = new ArrayList<>();
    private List<String> titles = new ArrayList<>();
    private ViewPager viewPager;
    private TabLayout tab;
    @Override
    public void onCreate(Bundle savedInstanceState) {
        super.onCreate(savedInstanceState);
    }
    @Override
    public View onCreateView(LayoutInflater inflater,ViewGroup container,
    Bundle savedInstanceState) {
        return inflater.inflate(R.layout.fragment_study,container,false);
    }
    @Override
    public void onViewCreated(View view,Bundle savedInstanceState) {
        super.onViewCreated(view,savedInstanceState);
        //初始化 Viewpager
        initViewPager(view);
        //初始化 TabLayout
        initTabLayout(view);
    }
    //初始化 ViewPager
    private void initViewPager(View view) {
        viewPager = view.findViewById(R.id.vp_study);
        fragments.add(new StudySubFragment1());
        fragments.add(new PassSubpage2Fragment());
        fragments.add(new PassSubpage2Fragment());
        ViewPagerAdapter adapter = new ViewPagerAdapter(getFragmentManager(),
        fragments,titles);
        viewPager.setAdapter(adapter);
    }
    //初始化 TabLayout
    private void initTabLayout(View view) {
        tab = view.findViewById(R.id.tab_study);
        titles.add("时政要闻");
        titles.add("科技发展");
        titles.add("技术技能");
        tab.setupWithViewPager(viewPager);
    }}
```

任务反思

编写并运行程序,将在代码编写及程序调试过程中出现的异常信息、产生原因及解决方法记录在下方。

问题 1：_____

产生原因：_____

_____

解决方法：_____

_____

问题 2：_____

产生原因：_____

_____

解决方法：_____

任务总结及巩固

师兄：在这个任务中，我们用到了几个很好用的控件来实现更为丰富的界面功能。

小白：ViewPager 是实现滑动翻页效果的利器；TabLayout 能够方便地实现顶部导航，与 ViewPager 配合使用更是效果倍增；RecyclerView 让列表效果更为灵活丰富。这些控件使用起来更为方便，现在学习模块看起来挺像那么回事了！

师兄：这些控件并不是从一开始就有的，随着移动应用功能越来越强大，用户对功能及界面美观度需求越来越高，更多的控件被开发出来，方便开发者实现更为丰富的界面效果。而且控件不断推陈出新，比如 ViewPager 现在就已经推出了 ViewPager2，既解决了 ViewPager 中存在的 bug，也有了更为强大的功能。技术层出不穷，我们就要不断学习啊！

小白：那我回去研究研究 ViewPager2，看看与 ViewPager 有什么不同。

一、基础巩固

1. ViewPager 需要借助（　　）适配器为其提供显示的内容 View。

   A. BaseAdapter　　　　　　　　B. SimpleAdapter

   C. PagerAdapter　　　　　　　　D. ArrayAdapter

2. 下列方法中，（　　）可以为 ViewPager 设置事件监听。

   A. addOnPageChangeListener( )　　B. setOnPageChangeListener( )

   C. setOnPageSelectedListener( )　　D. addOnPageSelectedListener( )

3. （多选）顶部导航栏可以使用（　　）实现。

   A. TabLayout　　　　　　　　　B. Toolbar

   C. SearchView　　　　　　　　　D. LinearLayout + Button

4. 下列方法中，（　　）可以为 RecyclerView 设置布局管理器。
  A. setAdatper( )　　　　　　　　　　B. setLayoutManager( )
  C. getItemCount( )　　　　　　　　　D. addItemDecoration( )
5. （多选）在使用 RecyclerView 时，可以使用（　　）布局管理器。
  A. LinearLayoutManager　　　　　　B. GridLayoutManager
  C. StaggeredGridLayoutManager　　　D. RelativeLayoutManager

二、技术实践

根据本任务学习的内容，完成图 3-2 中的科技发展列表及图 3-3 中的电台页面。

任务3　通过网络获取资讯数据

任务描述

通过网络获取服务器端传来的时政学习内容，进行数据的解析，并显示在时政列表中。

任务学习目标

通过本任务需达到以下目标：
➢ 能够通过 HTTP 访问网络。
➢ 能够使用 OkHttp 框架实现网络通信。
➢ 能够理解 JSON 数据格式并对 JSON 数据进行解析。

技术储备

一、通过 HTTP 访问网络

在移动互联网时代，手机已经成为人们生活中不可缺少的工具。应用程序中的很多数据都是通过网络从服务器端获得的。网络通信中最基本的协议为 HTTP，下面我们从 HTTP 开始对 Android 网络编程知识进行介绍。

（一）HTTP 通信

"有问题问百度"是我们学习软件开发过程中的"经验"，开发过程中遇到的各种问题，通过百度搜索往往总能找到解决方法。访问百度的过程就是通过 HTTP 完成的。所谓的 HTTP（Hyper Text Transfer Protocol，超文本传输协议）规定了浏览器和服务器之间互相通信的规则。

HTTP 是一种请求/响应式的协议，当客户端与服务器端建立连接后，向服务器端发送的请求，被称作"HTTP 请求"。服务器端接收到请求后会做出响应，称为"HTTP 响应"。其通信过程如图 3-18 所示。

图 3-18　HTTP 请求与响应

当我们通过手机浏览器也就是客户端访问百度时，会发送一个 HTTP 请求，当服务器端接收到这个请求后，会做出响应并将百度页面（数据）返回给客户端浏览器，这个请求和响应的过程实际上就是 HTTP 通信的过程。

（二）使用 HttpURLConnection 访问网络

Android 对 HTTP 通信提供了很好的支持，通过标准 Java 类 HttpURLConnection 便可实现基于 URL 的请求及响应功能。HttpURLConnection 继承自 URLConnection 类，它可以发送和接收任何类型和长度的数据，也可以设置请求方式、超时时间，具体用法如下。

```
1   HttpURLConnection connection = null;
2   try {
3       //(1)在 URL 的构造方法中传入要访问资源的路径
4       URL url = new URL("http://www.baidu.com");
5       //(2)获取连接对象
6       connection = (HttpURLConnection) url.openConnection();
7       //(3)设置请求方式
8       connection.setRequestMethod("GET");
9       //(4)设置连接超时毫秒数
10      connection.setConnectTimeout(8000);
11      //(5)获取服务器返回的输入流
12      InputStream in = connection.getInputStream();
13      //(6)读取数据内容
```

```
14      } catch (Exception e) {
15          e.printStackTrace();
16          return e.getMessage();
17      } finally {
18          //(7)连接关闭
19          if (connection ! = null) {
20              connection.disconnect();
21          }
22      }
```

上述代码中,通过从创建 URL 对象到关闭连接的 7 个步骤实现了客户端与服务器建立连接并获取服务器返回数据的过程。需要注意的是,在使用 HttpURLConnection 对象访问网络时,需要设置超时时间,以防止连接被阻塞时无响应,影响用户体验。

在第 8 行代码中,通过 setRequestMethod( )方法设置了网络请求方式。在使用 HttpURL-Connection 访问网络时,通常会用到两种网络请求方式:一种是 GET,一种是 POST。这两种请求方式用于表明请求指定资源的不同操作方式。下面分别对 GET 方式提交数据和 POST方式提交数据进行介绍。

1. GET 方式提交数据

使用 GET 方式表示希望从服务器获取数据。在获取数据时,有时也需要向服务器提交比如用户名、密码等参数,这些参数直接跟在请求 URL 后面。使用 GET 方式提交数据的具体代码如下。

```
1   String path = "https://image.baidu.com/search/index? " +
2           "tn = baiduimage&ps = 1&ct = 201326592&lm = -1&cl = 2&nc = 1&" +
3           "ie = utf - 8&word = HttpURLConnection";
4   URL url = new URL(path);          //设置 URL 对象
5   HttpURLConnection conn = (HttpURLConnection)url.openconnection();
6   conn.setRequestMethod("GET");//设置请求方式
7   conn.setConnectTimeout (5000);//设置超时时间
8   int responseCode = conn.getResponseCode();//获取状态码
9   if (responseCode = = 200){  //访问成功
10      Inputstream is = con.getInputstream();//获取服务器返回的输入流
11  }
```

在上述代码中,定义了一个网络请求路径,这是在百度中搜索 HttpURLConnection 图片的资源路径。其中,"?"之前的部分是网址,之后用"&"拼接了多个搜索条件,这里我们不去深究这些搜索条件都是什么,但可以看出最后一个"word"就是我们在搜索框中输入的搜索内容。

2. POST 方式提交数据

使用 POST 方式表示希望提交数据给服务器。提交的数据是以键值对的形式封装在请求实体中,用户通过浏览器无法看到发送的请求数据,因此,POST 方式要比 GET 方式相对安全。以 POST 方式提交数据的前半部分与 GET 方式类似，同样需要设置 URL、请求方式、超时时间，不同的地方有以下几点。

（1）POST 方式需要通过 setRequestProperty() 设置请求属性，该方法接收的参数为 <key, value>，key 代表属性名，value 代表属性值。

（2）POST 方式提交的数据通过流的形式写到服务器上。

具体代码如下。

```
1   String path = "https://image.baidu.com/search/index";
2   URL url = new URL(path);                          //设置 URL 对象
3   HttpURLConnection conn = (HttpURLConnection)url.openconnection();
4   conn.setRequestMethod("POST");                    //设置请求方式
5   conn.setConnectTimeout(5000);                     //设置超时时间
6   //封装要提交的数据,用 & 连接
7   String data = "tn = baiduimage&ps = 1&ct = 201326592&lm = -1&cl =
8                  2&nc = 1&" + "ie = utf - 8&word = HttpURLConnection";
9   //设置请求属性 Content - Type 的值,用于指定提交的实体数据的内容类型
10  conn.setRequestProperty("Content - Type",
11                 "application/x - www - form - urlencoded");
12  //设置请求属性 Content - Length 的值为提交数据的长度
13  conn.setRequestProperty("Content - Length",data.length() + "");
14  conn.setDoOutput(true);                           //设置允许向外写数据
15  OutputStream os = conn.getOutputStream();         //利用输出流往服务器写数据
16  os.write(data.getBytes());
17  int responseCode = conn.getResponseCode();        //获取状态码
18  if (response Code = =200){                        //访问成功
19      Inputstream is = con.getInputstream();//获取服务器返回的输入流
20  }
```

上述代码中，第 7~8 行为需提交的数据；第 9~13 行设置请求属性 Content - Type 和 Content - Length；第 14~15 行获取输出流对象并将数据写出。

【案例 3-1】 实现对百度网站的网络访问。

1. 案例分析

在本案例界面中，添加 1 个 Button 控件，单击按钮建立与百度的网络连接，并进行数据读取，将数据显示在界面上。在实际应用中，一般会采用异步消息处理进行网络连接。在子线程中进行网络连接，将返回的数据放入 Message 中，通过 Handler 返回给主线程进行界面更新。

2. 实现步骤

（1）创建名为"HttpDemo"的项目，并实现项目的界面布局。

界面布局非常简单，包含一个 Button 和一个 TextView，具体代码此处不再列出。

（2）创建名为"HttpUtil"的工具类，创建 sendHttpRequest() 方法实现网络连接及数据读取，具体代码如下。

```java
1   public class HttpUtil {
2       public static String sendHttpRequest(String address) {
3           HttpURLConnection connection = null;
4           try {
5               URL url = new URL(address);
6               //获取连接对象
7               connection = (HttpURLConnection) url.openConnection();
8               //设置请求方式
9               connection.setRequestMethod("GET");
10              //设置连接超时毫秒数
11              connection.setConnectTimeout(8000);
12              //设置读取超时时间
13              connection.setReadTimeout(8000);
14              //设置连接可以用于输入,默认true
15              connection.setDoInput(true);
16              //使用流对象进行网络数据的读取
17              InputStream in = connection.getInputStream();
18              BufferedReader reader = new BufferedReader(
19                              (new InputStreamReader(in)));
20              StringBuilder response = new StringBuilder();
21              String line;
22              while ((line = reader.readLine()) ! = null) {
23                  response.append(line);
24
25              }
26              return response.toString();
27          } catch (Exception e) {
28              e.printStackTrace();
29              return e.getMessage();
30          } finally {
31              //连接关闭
32              if (connection ! = null) {
33                  connection.disconnect();
34              }
35          }
36      }
37  }
```

（3）在 MainActivity.java 中，编写功能逻辑代码。

根据案例要求，需要在单击按钮时创建子线程实现网络连接，并创建 Handler 对象，重写 handleMessage() 方法进行界面更新。

```java
1   public class MainActivity extends AppCompatActivity {
2       private TextView textView;
3       private Button button;
4       //创建一个 Hanlder 对象,消息的发送和处理
```

```java
5    private Handler handler = new Handler(){
6        @Override
7        public void handleMessage(@NonNull Message msg) {
8            super.handleMessage(msg);
9            switch (msg.what){
10               case 1:
11                   textView.setText((String)msg.obj);
12           }
13       }
14   }
15   @Override
16   protected void onCreate(Bundle savedInstanceState) {
17       super.onCreate(savedInstanceState);
18       setContentView(R.layout.activity_main);
19       button = findViewById(R.id.send_request);
20       textView = findViewById(R.id.respont_text);
21       button.setOnClickListener(new View.OnClickListener() {
22           @Override
23           public void onClick(View v) {
24               //创建并启动线程
25               new Thread(new Runnable() {
26                   @Override
27                   public void run() {
28                       String address = "http://www.baidu.com";
29                       String response = HttpUtil.sendHttpRequest(address);
30                       //发送消息
31                       Message msg = new Message();
32                       msg.what = 1;
33                       msg.obj = response;
34                       //发送消息
35                       handler.sendMessage(msg);
36                   }
37               }).start();
38           }
39       });
40   }
41 }
```

在上述代码中，第 23~40 行的按钮单击事件处理方法 onClick() 中，创建并启动了子线程，在 run() 方法中调用 HttpUtil 工具类的 sendHttpRequest() 方法建立网络连接，并返回读取的数据；创建 Message 对象，将返回的数据放入 obj 字段，调用 Handler 的 sendMessage() 方法发送消息。

第 7~14 行的 handleMessage() 方法中，获取 Message 的 obj 字段，并将数据在页面上更新。

（4）在 AndroidManifest.xml 中，加入网络访问权限。

在 application 标签的外面，加入如下静态权限申请语句，否则，在运行时会出现异常。

```
<uses-permission android:name="android.permission.INTERNET"/>
```

运行程序，我们会看到如图 3-19 所示的运行效果。程序中我们访问了"www.baidu.com"网址，但程序运行结果并不是我们通常看到的百度页面，而是一大段代码。这是由于百度页面本身就是我们现在看到的 HTML 代码的形式，是浏览器将代码解析为百度页面。

图 3-19 运行效果

二、使用 OkHttp 框架访问网络

前面我们学习了 Android 原生提供的网络请求——HttpURLConnection。除了这种网络请求方式外，还有很多第三方网络请求框架可以提供方便、高效的网络请求及处理方法，比如OkHttp、Volley、Retrofit 等。下面我们通过 OkHttp 的使用感受网络请求框架的便捷。

OkHttp 是一款优秀的处理网络请求的开源框架，它支持 GET 请求和 POST 请求，基于Http 的文件上传和下载，加载图片支持响应缓存，避免重复的网络请求，支持使用连接池来降低响应延迟问题。

在使用 OkHttp 之前，需要注入依赖，打开项目的 build.gradle 文件，在 dependencies 中加入，具体代码如下。

```
implementation'com.squareup.okhttp3:okhttp: 4.9.1'
```

下面我们详细介绍如何使用 OkHttp 实现 GET 请求和 POST 请求。

(一) GET 请求

使用 OkHttp 框架发送 GET 请求有如下 4 个步骤。

(1) 创建 OkHttpClient 对象。
(2) 构造 Request 请求对象。
(3) 调用 OkHttpClient 的 newCall()方法创建 Call 对象。
(4) 调用 Call 对象的 execute()或 enqueue()方法实现同步或异步请求。

具体代码如下所示。

1. 同步方法

```
1   String url = "https://www.baidu.com";
2   OkHttpClient okHttpClient = new OkHttpClient();
3   final Request request = new Request.Builder()
4           .url(url)
5           .build();
6   final Call call = okHttpClient.newCall(request);
7   new Thread(new Runnable() {
8       @Override
9       public void run() {
10          try {
11              Response response = call.execute();
12              Log.d(TAG,"run: " + response.body().string());
13          } catch (IOException e) {
14              e.printStackTrace();
15          }
16      }
17  }).start();
```

上述代码中，第 3~5 行构造 Request 请求对象，传入访问资源路径 url；第 6 行创建 Call 对象。由于同步方法会阻塞调用线程，因此第 7~17 行创建并启动了子线程，在子线程中调用 execute()方法实现同步请求。

2. 异步方法

```
1   String url = "https://www.baidu.com";
2   OkHttpClient okHttpClient = new OkHttpClient();
3   final Request request = new Request.Builder()
4           .url(url)
5           .get()//默认就是 GET 请求,可以不写
6           .build();
7   Call call = okHttpClient.newCall(request);
8   call.enqueue(new Callback() {
```

```
9       @Override
10      public void onFailure(Call call,IOException e){
11          Log.d(TAG,"onFailure: ");
12      }
13
14      @Override
15      public void onResponse(Call call,Response response) throws IOException{
16          Log.d(TAG,"onResponse: " + response.body().string());
17      }
18  });
```

上述代码中，第 8~18 行调用 enqueue( ) 方法，实现接口 Callback 中的 onFailure( ) 和 onResponse( ) 方法，分别在请求失败和请求成功时调用，onResponse( ) 方法的第 2 个参数为 Response 对象，通过该对象可以获得返回数据。

（二）POST 请求

POST 方式与 GET 方式的区别在于，在构造 Request 对象时，POST 方式需要多构造一个 RequestBody 对象，用它来携带要提交的数据。在构造 RequestBody 对象时需要指定 MediaType，用于描述请求/响应 body 的内容类型。

下面两段代码分别提交了 JSON 数据和键值对。

1. 提交 JSON 数据

```
1   MediaType type = MediaType.parse("application/json;charset=utf-8");
2   OkHttpClient client = new OkHttpClient();
3   Public String post(String url,String json) throws IOException{
4       RequestBody body = RequestBody.create(type,json);
5       Request request = new Request.Builder()
6           .url(url)
7           .post(body)
8           .build();
9       client.newCall(request).enqueue(new Callback(){
10          @Override
11          public void onFailure(Request request,IOException e){
12          }
13          @Override
14          public void onResponse(Response response) throws IOException{
15              Log.d("json提交",response.body().string());
16          }
17      });
18  }
```

2. 提交键值对

```
1   OkHttpClient client = new OkHttpClient();
2   String post(String url) throws IOException{
3       RequestBody formBody = new FormEncodingBuilder()
```

```
4          .add("username","lujing")
5          .add("password","1234567")
6          .build();
7      Request request = new Request.Builder()
8           .url(url)
9           .post(body)
10          .build();
11     client.newCall(request).enqueue(new Callback() {
12          @Override
13          public void onFailure(Request request,IOException e) {
14          }
15          @Override
16          public void onResponse(Response response) throws IOException {
17               Log.d("键值对提交",response.body().string());
18          }
19     });
20  }
```

【案例 3 – 2】 使用 OkHttp 实现对百度网站的网络访问。

1. 案例分析

本案例界面与【案例 3 – 1】相同，案例实现中，采用异步方式进行网络访问。

2. 实现步骤

（1）创建名为"OkHttpDemo"的项目，并实现项目的界面布局。

（2）创建名为"HttpUtil"的工具类，创建 sendOkHttpRequest() 方法实现网络连接及数据读取，具体代码如下。

```
1  public class HttpUtil {
2      public static void sendOkHttpRequest(String address,Callback callback){
3          OkHttpClient client = new OkHttpClient();
4          Request request = new Request.Builder()
5                  .url(address)
6                  .build();
7          client.newCall(request).enqueue(callback);
8      }
9  }
```

sendOkHttpRequest() 中定义了两个参数，分别为网络访问地址和 Callback 对象。

（3）在 MainActivity.java 中，编写功能逻辑代码。

根据案例要求，需要在单击按钮时调用 sendOkHttpRequest() 方法实现网络连接，并在 onResponse() 方法中获取数据更新在界面上，具体代码如下。

```
1  public class MainActivity extends AppCompatActivity {
2      private TextView textView;
3      private Button button;
4      private String address;
```

```
5       private String responseData;
6
7       @Override
8       protected void onCreate(Bundle savedInstanceState) {
9           super.onCreate(savedInstanceState);
10          setContentView(R.layout.activity_main);
11          button = findViewById(R.id.send_request);
12          textView = findViewById(R.id.respont_text);
13          address = "https://www.baidu.com";
14          button.setOnClickListener(new View.OnClickListener() {
15              @Override
16              public void onClick(View view) {
17                  HttpUtil.sendOkHttpRequest(address,new Callback(){
18                      @Override
19                      public void onFailure(Call call,IOException e) {
20                          Log.e("okhttp","连接失败");
21                      }
22                      @Override
23                      public void onResponse(Call call,Response response)
24                                              throws IOException {
25                          //得到服务器返回的具体内容
26                          Log.e("okhttp","连接成功");
27                          responseData = response.body().string();
28                          runOnUiThread(new Runnable() {
29                              @Override
30                              public void run() {
31                                  textView.setText(responseData);
32                              }
33                          });
34                      }
35                  });
36              }
37          });
38      }
39  }
```

上述代码中，第28~33行，调用了runOnUiThread()方法回到主线程进行界面更新。

三、JSON 数据解析

通过网络请求从服务器上获取到的数据大部分是JSON类型，这种格式的数据不能够直接显示到界面上，需要将该数据解析为一个集合或对象的形式才可以显示到界面上，下面将针对 JSON 数据及其解析进行详细讲解。

（一）JSON 数据

JSON（JavaScript Object Notation，对象表示法）是一种轻量级的数据交互格式。它采用完全独立于编程语言的文本格式来存储和表示数据，简洁和清晰的层次结构使得 JSON 成为理想的数据交换语言。JSON 数据易于阅读和编写，同时也易于机器解析和生成，能够有效地提升网络传输效率。

JSON 的编码格式为"utf-8"，可以传输一个简单的数据，如 String、Number、Boolean，也可以传输一个数组或者一个复杂的对象。JSON 数据有两种表示结构，分别是对象结构和数组结构。

1. 对象结构

对象结构以"{"开始，以"}"结束。中间部分由以","分隔的键值对（key: value）构成，最后一个键值对后不用加","，键（key）和值（value）之间以":"分隔，其语法结构如下所示。

```
{
    key1:value1,
    key2:value2,
    ……
}
```

上述语法结构中的 key1、key2 必须为 String 类型，value1、value2 可以是 String、Integer、Object、Array 等数据类型。例如，一个 People 对象包含姓名、年龄、性别等信息，JSON 的表现形式如下。

```
{
    "name":"smith",
    "age":30,
    "sex":"男"
}
```

JSON 对象的属性也可以是一个对象，在下面的代码中，学校信息就是一个对象，其中包含学校名称和学校地址两个信息。

```
{
    "name":"smith",
    "age":28,
    "sex":"男",
    "school":{
        "sname":"威海职业学院",
        "address":"威海市科技新城"
    }
}
```

2. 数组结构

数组结构以"["开始，以"]"结束。中间部分由 0 个或多个以","分隔的对象（value）的列表组成，其语法结构如下所示。

```
[
    value1,
    value2,
```

```
    ……
]
```

其中，value 代表一个对象，该对象可以是具体的数据（如"4""您好"），也可以是一个对象结构的 JSON 数据。

（二）JSON 解析

从网络获取的 JSON 数据需要经过解析才能够进行进一步的处理或显示。假设现在有两条 JSON 数据，其中，json1 是对象结构的数据，json2 是数组结构的数据，示例代码如下。

json1：

```
{"name":"smith","age":30,"married":"married"}
```

json2：

```
[
    {"name":"smith","empno":1001,"job":"clerck","sal":9000.00},
    {"name":" Tom ","empno":1002,"job":"clerck","sal":9000.00},
    {"name":" Jack ","empno":1003,"job":"clerck","sal":9000.00},
]
```

1. 使用 JSONObject 与 JSONArray 类解析 JSON 数据

为了解析 JSON 数据，Android SDK 为开发者提供了 org.json 包，该包存放了解析 JSON 数据的类，其中，最重要的两个类是 JSONObject 和 JSONArray，JSONObject 用于解析对象结构的 JSON 数据，JSONArray 用于解析数组结构的 JSON 数据。

（1）使用 JSONObject 类解析对象结构的 JSON 数据，具体代码如下。

```
1    JSONObject jsonObject = new JSONObject(json1);
2    String name = jsonObject.optString("name");
3    int age = jsonObject.optInt("age");
4    boolean married = jsonObject.optBoolean("married");
```

上述代码中，首先创建了 JSONObject 类的对象 jsonObject，JSONObject 构造方法中传递的参数是对象结构的 JSON 数据，接着分别通过 jsonObject 的 optString()方法、optInt()方法、optBoolean()方法获取 JSON 数据中的 String 类型、int 类型、boolean 类型的数据。

（2）使用 JSONArray 类解析数组结构的 JSON 数据，具体代码如下。

```
1    JSONArray jsonArray = new JSONArray(json2);
2    for(int i = 0;i < jsonArray.length();i++){
3        JSONObject jsonObject = jsonArray.getJSONObject(i);
4        String name = jsonObject.optString("name");
5        String empno = jsonObject.optString("empno");
6        String job = jsonObject.optString("job");
7        double sal = jsonObject.optDouble(sal);
8    }
```

上述代码中，首先创建了 JSONArray 类的对象 jsonArray，JSONArray 构造方法中传递的参数是数组结构的 JSON 数据，接着通过一个 for 循环来遍历 JSONArray 中的数据。因为 json2 是一个数组结构的数据，所以需要在 for 循环中对数组中的数据进行遍历。在 for 循环中，首先需要通过 getJSONObject() 方法获取数组中的每个对象，接着通过该对象的 optString() 与 optDouble() 方法获取 json2 中对应的数据。

OptXXX() 在解析数据时，如果对应的字段不存在，这些方法会有默认的返回值。

2. 使用 Gson 库解析 JSON 数据

为了更便捷地解析 JSON 数据，Google 提供了一个 Gson 库，该库中定义了 fromJson() 方法来解析 JSON 数据。使用 Gson 库必须先导入依赖 "com.google.code.gson：gson：2.8.7"。

使用 Gson 库之前必须创建 JSON 数据对应的实体类，实体类中的成员变量名必须与 JSON 数据中的 key 值一致。

下面我们针对 json1 和 json2 采用 Gson 库进行数据解析。

(1) 创建 JSON 数据对应的实体类。

实体类中的成员名称必须与 JSON 数据中的 key 值一致。

json1 及 json2 对应的实体类代码如下。

```
1   //json1 对应实体类
2   class Person1{
3       private String name;
4       private int age;
5       private boolean married;
6       //省略构造方法及 getter、setter 方法
7   }
8   //json2 对应实体类
9   class Person2{
10      private String name;
11      private intempno;
12      private String job;
13      privatedouble sal;
14      //省略构造方法及 getter、setter 方法
15  }
```

(2) 使用 Gson 库解析对象结构的 JSON 数据，具体代码如下。

```
1   Gson gson = new Gson();
2   Person1 person1 = gson.fromJson(json1,Person1.class);
```

(3) 使用 Gson 库解析数组结构的 JSON 数据，具体代码如下。

```
1   Gson gson = new Gson();
2   Type listType = new TypeToken<List<Person2>>(){}.getType();
3   List<Person2> person2 = gson.fromJson(json2,listType);
```

通过两种解析方式的对比可以看出，Gson 库解析 JSON 数据的代码简单快捷，便于提高开发效率。

任务实施

1. 任务分析

本任务需模拟从服务器端获取 JSON 格式的时政学习信息,将 JSON 数据进行解析并显示在 RecyclerView 列表中。

2. 实现步骤

我们在任务2的基础上,继续完成本任务。

(1) 读取模拟服务器端的时政学习 JSON 数据。

本任务需要读取的 JSON 数据如下所示。

```
1  [
2    {
3      "title":"工信部:到2025年基本建成安全可靠的新型数据基础设施",
4      "content":"工信部:到2025年基本建成安全可靠的新型数据基础设施",
5      "date":"2021-07-13",
6      "source":""学习通关"学习平台"
7    },
8    {
9      "title":"第二届国际人工智能算例性能榜出炉"鹏城云脑Ⅱ"蝉联榜首",
10     "content":"第二届国际人工智能算例性能榜出炉"鹏城云脑Ⅱ"蝉联榜首",
11     "date":"2021-07-13",
12     "source":""学习通关"学习平台"
13   },
14   {
15     "title":"牢牢把握中华民族伟大复兴这一主题学习时报",
16     "content":"牢牢把握中华民族伟大复兴这一主题学习时报",
17     "date":"2021-07-13",
18     "source":""学习通关"学习平台"
19   }
20 ]
```

可以自行安装 tomcat 服务器作为模拟服务器,也可以通过一些模拟数据网站进行服务器端数据的模拟。比如,fastMock 就是一个在线数据接口模拟平台。在该平台上注册→创建项目→新增接口→放入 JSON 数据,即可获得数据接口链接。

(2) 新建 util 包,在包中创建 HttpUtil 工具类。

工具类中包含两个方法:sendRequestWithOkHttp()用于创建网络连接,parseJSONWithGSON()用于解析 JSON 数据。

【代码3-10】 HttpUtil.java

```
1  public class HttpUtil {
2
3      public static void sendRequestWithOkHttp(String address,
4                                               Callback callback) {
```

```
5        OkHttpClient client = new OkHttpClient();
6        Request request = new Request.Builder()
7               .url(address)
8               .build();
9        client.newCall(request).enqueue(callback);
10    }
11    public static List parseJSONWithGSON(String jsonData){
12       List<News> list;
13       Gson gson = new Gson();
14       Type listType = new TypeToken<List<News>>(){}.getType();
15       list = gson.fromJson(jsonData,listType);
16       return list;
17    }
18 }
```

(3) 在StudySubFragment1.java中,修改功能逻辑代码。

首先,数据将通过资源路径"https://www.fastmock.site/mock/22aeadc4d44b088ae36dd7c00bd3ad57/learntopass/currentnews"(此路径来自fastMock平台创建的数据接口)获取,可以去掉其中的initData()方法,改由调用工具类中的方法进行网络连接及数据解析获取显示数据集合。

【代码3-11】 StudySubFragment1.java

```
1  public class StudySubFragment1 extends Fragment{
2     List<News> newsList = new ArrayList<>();
3     public static final String ADDRESS = "https://www.fastmock.site/mock/" +
4         "22aeadc4d44b088ae36dd7c00bd3ad57/learntopass/currentnews";
5     //省略onCreate()方法及onCreateView()方法
6     @Override
7     public void onViewCreated(final View view,Bundle savedInstanceState){
8        super.onViewCreated(view,savedInstanceState);
9        HttpUtil.sendRequestWithOkHttp(ADDRESS,new Callback(){
10          @Override
11          public void onFailure(Call call,IOException e){
12             Log.e("okhttp","连接失败");
13          }
14          @Override
15          public void onResponse(Call call,Response response)
16                                   throws IOException{
17             Log.e("okhttp","连接成功");
18             String responseData = response.body().string();
19             newsList = HttpUtil.parseJSONWithGSON(responseData);
20             //回到主线程初始化RecyclerView
21             getActivity().runOnUiThread(new Runnable(){
22                @Override
```

```
23                  public void run() {
24                      initRecyclerView(view);
25                  }
26              });
27          }
28      });
29  }
30  //省略 initRecyclerView()方法
31 }
```

🌀 任务反思

编写并运行程序，将在代码编写及程序调试过程中出现的异常信息、产生原因及解决方法记录在下方。

问题1：

产生原因：

解决方法：

问题2：

产生原因：

解决方法：

🌀 任务总结及巩固

小白：网络通信感觉有点复杂，但是 OkHttp 框架让整个过程简单了许多，Gson 也是如此。

师兄：是的，为了让开发更为便捷，Android 中有不少特定功能的框架，比如网络框架 Retrofit、使用注解生成模板代码的 ButterKnife、图表框架 MPAndroidChart 等。它们都对原生 Android 代码进行了封装，能够提供强大的功能实现。

小白：看来要学的东西还有很多呀。

师兄：那当然啦。随着 Android 版本的不断更新，新的技术、新的用法、新的框架也层出不穷，唯有不断学习才能跟上技术进步的脚步。再给你个自学任务：图片是我们经常需要从服务器上获得的数据，去找找资料，看如何从服务器下载图片？

一、基础巩固

1. 在使用 HttpURLConnection 进行网络通信时，（　　）方法可以设置请求方式。
   A. openConnection( )　　　　　　　　B. setRequestMethod( )
   C. setConnectTimeout( )　　　　　　　D. setRequestProperty( )
2. （多选）下列框架中，（　　）是网络请求框架。
   A. Butterknife　　　B. OkHttp　　　C. Volley　　　D. Gson
3. （多选）OkHttp 框架中（　　）方法可以实现同步或异步请求。
   A. newCall( )　　　B. execute( )　　　C. enqueue( )　　　D. onResponse( )
4. （多选）JSON 数据主要有（　　）两种表示结构。
   A. 基本类型结构　　　B. 对象结构　　　C. 数组结构　　　D. 集合结构
5. 使用 Gson 解析 JSON 数据，以下说法不正确的是（　　）。
   A. 使用 Gson 库不需要导入依赖
   B. 使用 Gson 库之前需要创建 JSON 对应的实体类
   C. 可以调用 fromJson( ) 方法解析对象结构数据
   D. 在解析数组结构的 JSON 数据时，先创建 TypeToken

二、技术实践

利用本任务学习的内容，对学习模块的其他学习内容相关代码进行改写，实现通过网络读取并解析显示数据。

学习目标达成度评价

了解对本模块的自我学习情况，完成表 3-2。

表 3-2　学习目标达成度评价表

序号	学习目标	学生自评
1	能够使用 ToolBar 和 BottomNavigationView 实现顶部工具栏及底部导航功能	□能够熟练实现 □通过查看以往代码及课本，能够基本实现 □不会实现
2	能够使用 TabLayout 及 ViewPager 控件实现顶部标签栏及下方滑动翻页效果	□能够熟练实现 □通过查看以往代码及课本，能够基本实现 □不会实现
3	能够使用 RecyclerView 实现列表效果	□能够熟练实现 □通过查看以往代码及课本，能够基本实现 □不会实现
4	能够理解通过 HttpURLConnection 访问网络的方法且能够运用 OkHttp 框架实现网络通信	□能够实现网络通信 □通过查看以往代码及课本，能够基本实现 □不会实现
5	能够理解 JSON 数据并实现 JSON 数据解析	□能够熟练实现数据解析 □通过查看以往代码及课本，能够基本实现 □不会实现
评价得分			
学生自评得分 （20%）	学习成果得分 （60%）	学习过程得分 （20%）	模块综合得分

（1）学生自评得分。

每个学习目标有 3 个选项，选择第 1 个选项得 100 分，选择第 2 个选项得 70 分，选择第 3 个选项得 50 分，学生自评得分为各项学习目标得分的平均分。

（2）学习成果得分。

教师根据学生阶段性测试及模块学习成果的完成情况酌情赋分，满分为 100 分。

（3）学习过程得分。

教师根据学生的其他学习过程表现，如到课情况、参与课程讨论等情况酌情赋分，满分为 100 分。

功能模块 4

电台模块的实现

> **说在前面**
>
> **小白**：师兄，通过前面 3 个功能模块的学习和实践，我的"学习通关"App 已经完成大部分功能了，还挺有成就感呢！
>
> **师兄**：可不是吗！虽然每个布局、每个控件、每种技术单独学习起来并没有多难，但放到项目中综合应用可没那么容易。前面几个模块你完成得非常不错，对于 Android 的基本开发也算是向前跨了一大步。
>
> **小白**：我们前期的设计中就差电台模块没有完成了。我有一个疑问，在前面的学习中，总是提到 Android 4 大组件，我们已经学了 Activity 和 ContentProvider 这两个，其余两个又是什么呢？
>
> **师兄**：别急，四大组件中的另外两个在电台模块中就会用到，而且在这部分我们还会学习视频与音频控件的应用。完成了这些，"学习通关"App 的基本功能就算具备了，咱们还得把项目打包，为发布上线做准备。所以，这部分我们将完成以下 3 个任务。
>
> 任务 1：视频及音频播放功能的实现。
>
> 任务 2：服务及广播的应用。
>
> 任务 3：项目打包。

功能需求描述

（1）单击"学习"界面下方的"电视台"导航菜单，进入如图 4－1 所示的视频列表；单击视频列表中的"播放"按钮，进入如图 4－2 所示的视频播放界面，进行视频播放。

（2）单击"学习"界面下方的"电台"导航菜单，进入如图 4－3 所示的音频列表；单击音频列表中的列表项，进入如图 4－4 所示的音频播放界面，能够通过后台服务播放音频。

图 4-1 视频列表　　图 4-2 视频播放界面　　图 4-3 音频列表　　图 4-4 音频播放界面

学习目标

1. 知识目标

(1) 视频控件 VideoView 的使用方法

(2) 音频控件 MediaPlayer 的使用方法

(3) 服务 Service 的使用方法

(4) 广播的使用方法

2. 技能目标

(1) 能够综合运用控件、布局及音视频控件实现音视频播放功能

(2) 能够通过 Service 实现音频的后台播放

(3) 能够发送和接收广播

任务1　视频及音频播放功能的实现

任务描述

根据前面的参考界面，完成图 4-1 界面中的视频列表，单击"播放"按钮跳转到该视频播放界面，界面效果如图 4-2 所示；完成图 4-3 界面中的音频列表，单击列表项弹出音频播放界面，界面效果如图 4-4 所示。

任务学习目标

通过本任务需达到以下目标：

➢ 能够使用 VideoView 控件播放视频。

➢ 能够使用 MediaPlayer 控件播放音频。

技术储备

一、VideoView 控件的应用

VideoView 是 Android 中提供的用于视频播放的控件，借助它可以完成一个简易的视频播放器。VideoView 提供了用于控制视频播放的方法，请通过查阅 API 文档学习这些方法的用法，并完成表 4-1。

查文档找答案

查找 API 文档，学习 VideoView 的继承关系及常用方法。

- VideoView 的父类是_____。
- VideoView 的子类有_____。

表 4-1 VideoView 的常用方法

方法名称	功能描述
setVideoPath( )	
start( )	
pause( )	
resume( )	
seekTo( )	
isPlaying( )	
getDuration( )	

通过 VideoView 控件播放视频的过程具体如下。

1. 在布局文件中添加 VideoView 控件

```
1  <VideoView
2      android:id = "@ + id/videoview"
3      android:layout_width = "match_parent"
4      android:layout_height = "match_parent" />
```

2. 设置视频路径并播放视频

```
1  VideoView videoview = (VideoView)findViewById(R.id.videoview);
2  videoview.setVideoPath("mnt/sdcard/video.mp4");//播放本地视频
3  videoview.setVideoURI(Uri.parse("http://www.xxx.com/video.mp4"));
4  //加载网络资源
5  videoview.start();//播放视频
```

上述代码中，setVideoPath( )用于播放本地视频，需要视频地址作为其参数。setVideoURI( )用于播放网络视频，通过调用 Uri 类的 parse( )方法将网络视频地址转换为 Uri 传递给 setVideoURI( )方法。

需要注意的是，播放网络视频时需要在 Manifest.xml 文件中的 <manifest> 标签中添加访问网络的权限，代码如下所示。

```
<uses-permission android:name="android.permission.INTERNET"/>
```

3. 为 VideoView 控件添加控制器

使用 VideoView 控件播放视频时，可以通过 setMediaController()方法为其添加一个控制器 MediaController 对象，该控制器中包含媒体播放器的一些典型按钮，如播放/暂停、倒带、快进、进度滑动器等。VideoView 控件能够绑定媒体播放器，从而使播放状态和控件中显示的图像同步，具体代码如下。

```
MediaController controller = new MediaController();
Videoview.setMediaController(controller);
```

【案例 4-1】 利用 VideoView 实现一个视频播放器，如图 4-5 所示。

1. 案例分析

本案例页面中包含一个 VideoView 控件及一个"播放/暂停"按钮，需在布局文件中加入控件。播放的资源文件放入 res 目录中的 raw 文件夹中。单击"播放/暂停"按钮，视频根据当前的状态播放或停止，同时，按钮的图片显示为对应的暂停或播放图片。

2. 实现步骤

（1）创建名为"VideoViewDemo"的应用程序。

（2）导入视频文件。在 res 文件夹中，创建 raw 文件夹，将视频文件 video.mp4 放入 raw 文件夹中。

（3）编写界面布局文件。在页面布局文件 activity_main.xml 中，放置 1 个 VideoView 控件用于显示视频，1 个 ImageView 控件用于显示"播放/暂停"按钮。

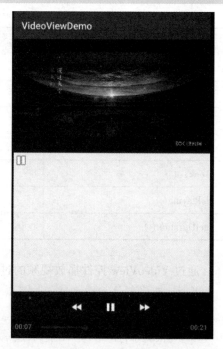

图 4-5 视频播放界面

```
1   <?xml version="1.0" encoding="utf-8"?>
2   <LinearLayout xmlns:android="http://schemas.android.com/apk/res/android"
3       xmlns:tools="http://schemas.android.com/tools"
4       android:layout_width="match_parent"
5       android:layout_height="match_parent"
6       android:orientation="vertical"
7       tools:context=".MainActivity">
8       <VideoView
9           android:id="@+id/video_view"
10          android:layout_width="match_parent"
```

```
11            android:layout_height = "300dp" />
12      < ImageView
13            android:id = "@ + id/bt_play"
14            android:layout_width = "wrap_content"
15            android:layout_height = "wrap_content" />
16   </LinearLayout >
```

（4）编写界面交互代码。在 MainActivity 中，创建一个 play( )方法，在该方法中实现视频播放功能，具体代码如下所示。

```
1   public class MainActivity extends AppCompatActivity
2                       implements View.OnClickListener {
3      private VideoView videoView;
4      private MediaController controller;
5      @Override
6      protected void onCreate(Bundle savedInstanceState) {
7         super.onCreate(savedInstanceState);
8         setContentView(R.layout.activity_main);
9         bt_play = (ImageView) findViewById(R.id.bt_play);
10        videoView = (VideoView) findViewById(R.id.video_view);
11        String url = "android.resource://" + getPackageName() +
12                     "/" + R.raw.fish;
13        Uri uri = Uri.parse(url);
14        videoView.setVideoURI(uri);
15        controller = new MediaController(this);
16        videoView.setMediaController(controller);
17        bt_play.setOnClickListener(this);
18     }
19     @Override
20     public void onClick(View v) {
21        switch (v.getId()) {
22           case R.id.bt_play:
23              play();
24              break;
25        }
26     }
27     //播放视频
28     private void play() {
29        if (videoView ! = null && videoView.isPlaying()) {
30           bt_play.setImageResource(android.R.drawable.ic_media_play);
31           videoView.stopPlayback();
32           return;
33        }
34        videoView.start();
35        bt_play.setImageResource(android.R.drawable.ic_media_pause);
36        videoView.setOnCompletionListener(
```

```
37                    new MediaPlayer.OnCompletionListener(){
38                @Override
39                public void onCompletion(MediaPlayer mp) {
40                    bt_play.setImageResource(android.R.drawable.ic_media_play);
41                }
42         });
43     }
44 }
```

第 11 ~ 14 行代码定义视频路径字符串，该路径指向 res/raw 目录的视频文件，转换为 Uri 对象，并通过 setVideoURI( ) 方法将视频文件的路径加载到 VideoView 控件上。

第 15 ~ 16 行代码通过 setMediaController( ) 方法为 VideoView 控件绑定控制器，该控制器可以显示视频的播放/暂停、快进、快退和进度条等按钮。

第 20 ~ 26 行代码实现了 onClick( ) 方法，单击下方按钮时调用 play( ) 方法播放视频。

第 28 ~ 43 行创建了 play( ) 方法，方法中首先通过 isPlaying( ) 方法判断当前是否正在播放，如果正在播放，则通过 setImageResource( ) 方法设置播放按钮图片为 "ic_media_play"，这个图片是 Android 中自带的图片；接着调用 stopPlayback( ) 方法停止播放视频；否则，调用 start( ) 方法播放视频，并设置暂停按钮的图片为 "ic_media_pause"，接着通过 setOnCompletionListener( ) 方法设置 VideoView 控件的监听器，当视频播放完时，会调用该监听器中的 onCompletion( ) 方法，在该方法中将播放按钮图片设置为 "ic_media_play"。

二、MediaPlayer 控件的应用

MediaPlayer 是 Android 中提供的用于音频播放的控件，借助它可以完成一个简易的音频播放器。MediaPlayer 提供了用于控制音频播放的方法，请通过查阅 API 文档学习它们的用法。

查文档找答案

查找 API 文档，学习 MediaPlayer 的常用方法，完成表 4 - 2。

表 4 - 2 MediaPlayer 的常用方法

方法名称	功能描述
create( )	
start( )	
pause( )	
prepare( )	
seekTo( )	
isPlaying( )	
stop( )	

MediaPlayer 的工作流程具体如下。

1. 创建 MediaPlayer 对象

使用 MediaPlayer 播放音频时，需创建一个 MediaPlayer 对象。

```
MediaPlayer mediaplayer = new MediaPlayer();     //第一种方法
MediaPlayer mediaplayer = MediaPlayer.create(this, R.raw.xxx);   //第二种方法
```

上面列出了两种创建 MediaPlayer 对象的方法，第一种方法直接通过 new 关键字创建对象；第二种方法调用 create( ) 方法创建对象，这种方法通常是通过 raw 文件夹中的音频文件创建，就不需要再调用 setDataSource( ) 方法了。

2. 设置音频路径

```
1  //设置应用自带音频文件
2  mediaplayer = MediaPlayer.create(MainActivity.this,R.raw.voice.mp4)
3  //存储在 SD 卡或其他文件路径下的媒体文件
4  mediaplayer.setDataSource("/sdcard/test.mp3");
5  //网络上的媒体文件
6  mediaplayer.setDataSource("http://www.citynorth.cn/music/confucius.mp3");
```

在播放来自网络的音频文件时，需要在 AndroidManifest.xml 文件中添加网络访问权限。

```
<uses-permission android:name="android.permission.INTERNET" />
```

3. 播放音频文件

调用 start( ) 方法可以播放音频文件，在此之前一般会调用 prepare( ) 方法或 prepareAsync( ) 方法将音频解析到内存中。prepare( ) 方法为同步操作，一般用于解析较小的文件，prepareAsync( ) 方法为异步操作，一般用于解析较大的文件，具体代码如下。

（1）播放小音频文件。

```
1  //调用 prepare( )方法进入准备状态
2  mediaplayer.prepare();
3  //调用 start( )方法播放音频
4  mediaplayer.start()
```

> **注意**：使用 create( ) 方法创建 MediaPlayer 对象并设置程序自带音频文件时，不需要调用 prepare( ) 方法，直接调用 start( ) 方法播放音频文件即可。

（2）播放大音频文件。

```
1  //调用 prepareAsync ()方法进入准备状态
2  mediaplayer. prepareAsync ();
3  mediaplayer.setOnPreparedListener(new OnPreparedListener){
4      public void onPrepared(MediaPlayer player){
5          player.start();
6      }
7  }
```

在上述代码中，prepareAsync()方法在子线程中异步操作，其执行不会影响主线程操作。setOnPreparedListener()方法用于设置 MediaPlayer 类的监听器，监听文件解析是否完成，如果完成，则调用 onPrepared() 方法。在 onPrepared() 方法中，调用 start() 方法播放音频文件。

4. 暂停播放

调用 pause() 方法可以暂停音频播放。在暂停播放之前，需要先判断 MediaPlayer 对象是否存在，且当前是否正在播放，具体代码如下。

```
1  If(mediaplayer! =null && mediaplayer.isPlaying()){
2      mediaplayer.pause();
3  }
```

5. 重新播放

调用 seekTo() 方法可以定位播放，可以用于快退或快进音频播放，该方法的参数表示将播放时间定位在多少毫秒，具体代码如下。

```
1  //在播放状态下进行重播
2  If(mediaplayer! =null && mediaplayer.isPlaying()){
3      mediaplayer.seekTo(0);//参数为0,表示从头开始播放音频
4  }
5  //在暂停状态下进行重播
6  If(mediaplayer! =null){
7      mediaplayer.seekTo(0);//参数为0,表示从头开始播放音频
8      mediaplayer.start();
9  }
```

6. 停止播放

调用 stop() 方法可以停止播放，停止播放后还需调用 release() 方法将 MediaPlayer 对象占用的资源释放，并将该对象设置为 null，具体代码如下。

```
<action android:name = "android.intent.action.MAIN" />
```

【案例 4-2】 利用 MediaPlayer 实现一个音乐播放器，如图 4-6 所示。

（1）创建程序。创建名为"MediaPlayerDemo"的应用程序。

（2）导入视频文件。将音频文件 test.mp3 放入"res/raw"目录中。

（3）编写界面布局文件。在页面布局文件 activity_main.xml 中，放置 3 个按钮，分别用于音频的播放、暂停和停止，一个拖动条控件 SeekBar 用来表示音频播放进度，两个 TextView 控件分别用来表示音频播放的时间和总时间。

图 4-6 音乐播放器

具体代码如下所示。

```xml
1  <LinearLayout xmlns:android="http://schemas.android.com/apk/res/android"
2      android:layout_width="match_parent"
3      android:layout_height="match_parent"
4      android:layout_margin="3dp"
5      android:orientation="vertical" >
6      <LinearLayout
7          android:layout_width="match_parent"
8          android:layout_height="wrap_content"
9          android:orientation="horizontal" >
10         <Button
11             android:id="@+id/play"
12             android:layout_width="0dp"
13             android:layout_height="wrap_content"
14             android:layout_weight="1"
15             android:text="播放" />
16
17         <Button
18             android:id="@+id/pause"
19             android:layout_width="0dp"
20             android:layout_height="wrap_content"
21             android:layout_weight="1"
22             android:text="暂停" />
23         <Button
24             android:id="@+id/stop"
25             android:layout_width="0dp"
26             android:layout_height="wrap_content"
27             android:layout_weight="1"
28             android:text="停止" />
29     </LinearLayout>
30     <SeekBar
31         android:id="@+id/seekbar"
32         android:layout_width="match_parent"
33         android:layout_height="wrap_content"
34         android:max="0"
35         android:progress="0"
36         android:secondaryProgress="0" />
37
38     <RelativeLayout
39         android:layout_width="match_parent"
40         android:layout_height="wrap_content" >
41         <TextView
42             android:id="@+id/tv"
43             android:layout_width="wrap_content"
44             android:layout_height="wrap_content"
45             android:layout_alignParentLeft="true"
46             android:text="当前时间" />
47         <TextView
48             android:id="@+id/tv2"
```

```
49                 android:layout_width = "wrap_content"
50                 android:layout_height = "wrap_content"
51                 android:layout_alignParentRight = "true"
52                 android:text = "总时间" />
53         </RelativeLayout>
54     </LinearLayout>
55
```

(4) 编写界面交互代码

在 MainActivity 中，实现音频播放相关功能。

```
1   public class MainActivity extends AppCompatActivity
2         implements View.OnClickListener,SeekBar.OnSeekBarChangeListener {
3      private Button play,pause,stop;
4      private MediaPlayer player;
5      private SeekBar mSeekBar;
6      private TextView tv,tv2;
7      private boolean hadDestroy = false;   //布尔型记录音频对象是否销毁
8      private Handler mHandler = new Handler();
9
10     Runnable = new Runnable() {
11        public void run() {
12           if (! hadDestroy) runnable {
13              mHandler.postDelayed(this,1000);
14              int currentTime = Math.round(player.
15                    getCurrentPosition()/1000);
16              String currentStr = String.format("% s% 02d:% 02d",
17                    "当前时间",currentTime/60,currentTime % 60);
18              tv.setText(currentStr);
19              mSeekBar.setProgress(player.getCurrentPosition());
20           }
21        }
22     };
23     @Override
24     protected void onCreate(Bundle savedInstanceState) {
25        super.onCreate(savedInstanceState);
26        setContentView(R.layout.activity_main);
27        play = (Button) findViewById(R.id.play);
28        pause = (Button) findViewById(R.id.pause);
29        stop = (Button) findViewById(R.id.stop);
30        mSeekBar = (SeekBar) findViewById(R.id.seekbar);
31        tv = (TextView) findViewById(R.id.tv);
32        tv2 = (TextView) findViewById(R.id.tv2);
33        mSeekBar.setOnSeekBarChangeListener(this);
34        play.setOnClickListener(this);
35        pause.setOnClickListener(this);
36        stop.setOnClickListener(this);
37        player = new MediaPlayer();
38        initMediaplayer();
39
40     }
```

```
41      @Override
42      public void onClick(View v){
43          switch(v.getId()){
44              case R.id.play://"播放"按钮被按下时执行
45                  if(!player.isPlaying()){
46                      player.start();//启动音频播放
47                      int totalTime=Math.round(
48                              player.getDuration()/1000);
49                      //计算音频播放总时间
50                      String str=String.format("% 02d:% 02d",totalTime/60,
51                              totalTime % 60);
52                      //将总时间转换为分秒形式显示
53                      tv2.setText(str);
54                      mSeekBar.setMax(player.getDuration());
55                      mHandler.postDelayed(runnable,1000);
56                  }
57                  break;
58              case R.id.pause://"暂停"按钮按下时执行
59                  if(player.isPlaying()){
60                      player.pause();//音频暂停播放
61                  }
62                  break;
63              case R.id.stop://按下"停止"按钮时执行
64                  if(player.isPlaying()){
65                      mSeekBar.setProgress(0);
66                      player.reset();
67                      initMediaplayer();
68                  }
69                  break;
70              default:
71                  break;
72          }
73      }
74      //初始化播放器
75      private void initMediaplayer(){
76          try{
77              player=MediaPlayer.create(MainActivity.this,R.raw.voice1);
78              player.prepare();
79          }catch(Exception e){
80              e.printStackTrace();
81          }
82      }
83      @Override
84      public void onProgressChanged(SeekBar seekBar,int progress,
85                      boolean fromUser){
86          if(player!=null){    //更新进度条显示进度
87              player.seekTo(seekBar.getProgress());
88          }
89      }
90      @Override
```

```
91      public void onStartTrackingTouch(SeekBar seekBar){
92          //TODO 自动生成的方法存根
93      }
94      @Override
95      public void onStopTrackingTouch(SeekBar seekBar){
96          //TODO 自动生成的方法存根
97      }
98      @Override
99      protected void onDestroy(){
100         //TODO 自动生成的方法存根
101         super.onDestroy();
102         if(player!=null){
103             player.stop();
104             hadDestroy=true;
105             player.release();
106             player=null;
107         }
108     }
109 }
```

第 3～22 行代码定义了本案例实现所需的变量，其中，Handler 对象 mHandler 用来把拖动条变化加入消息队列；runnable 对象在创建时重写了 run()方法，实现每隔 1 秒获取当前音频的时间并在界面上进行更新。

第 24～40 行的 onCreate()方法中获取控件对象并添加事件监听，其中，第 33 行为 SeekBar 控件添加 OnSeekBarChangeListener 事件监听，MainActivity 实现了 SeekBar.OnSeekBarChangeListener 接口，第 83～97 行重写了接口中的 3 个方法；第 83～89 行的 onProgressChanged()方法在拖动条被拖动时被调用，在方法中，调用 SeekBar 控件的 getProgress()方法获得当前拖动条的进度，并调用 MediaPlayer 对象的 seekTo()方法定位到当前进度，开始播放。

第 74～82 行代码定义了 initMediaplayer()方法，创建 MediaPlayer 对象，并调用 prepare()方法解析音频文件。

第 42～73 行实现了按钮单击事件的处理方法——onClick()方法，其中，第 44～57 行是单击"播放"按钮时的操作，调用 start()开始播放音频，调用 getDuration()方法获取音频时长，设置拖动条的最大值，并调用 Handler 对象的 postDelay()方法将对拖动条变化的操作加入消息队列；第 58～62 行是单击"暂停"按钮时的操作，调用 pause()方法暂停播放音频；第 63～68 行是单击"停止"按钮时的操作，重置 MediaPlayer 对象。

❖ 任务实施

1. 任务分析

本任务中视频功能需要在单击"学习"页面底部的"电视台"导航菜单时显示，因此，需要在 VideoFragment 中实现视频列表，列表中每项包含视频图片、视频标题等相关信息以及"播放"按钮，需要通过添加按钮单击事件监听，实现跳转至视频播放页面功能。

音频功能需要在单击"学习"页面底部的"电台"导航菜单时显示,因此,需要在 AudioFragment 中实现音频列表,列表页面主体也为 ListView 列表,列表项包含音频主要标题、音频时长等信息。需要通过添加列表项选中事件监听,实现单击音频列表项跳转至音频播放页面功能。

2. 实现步骤

我们在功能模块 3 的基础上继续实现本任务的相关功能。

(1) 设计并创建实体类。

根据本任务的功能需求,可以提取视频及音频实体类。视频实体类包括视频路径、视频图片、视频时长、视频标题等属性;音频实体类包括音频资源 ID、音频标题、音频内容、音频时长等属性。

【代码 4-1】 Video. java

```
1  public class Video {
2      private String uri;//播放视频路径
3      private int pic;//视频图片
4      private int time;//视频时长
5      private String title;//播放标题
6      //省略构造方法及 getter、setter 方法
7  }
```

【代码 4-2】 Audio. java

```
1  public class Audio implements Serializable {
2      private int rawId;//播放音频 ID
3      private String title;//播放标题
4      private String content;//音频内容
5      private int strTime;//音频时长
6      //省略构造方法及 getter、setter 方法
7  }
```

(2) 完成视频列表功能代码编写。

①编写 VideoFragment 的布局及视频列表的列表项布局。

VideoFragment 的布局文件为 "fragment_video.xml",布局中包含一个 ListView 控件;创建列表项布局文件 "video_item_list.xml",显示效果如图 4-7 所示。

图 4-7 列表项显示效果

【代码 4-3】 fragment_video. xml

```
1  <? xml version = "1.0" encoding = "utf-8"? >
2  <LinearLayout xmlns:android = "http://schemas.android.com/apk/res/android"
3      xmlns:tools = "http://schemas.android.com/tools"
4      android:layout_width = "match_parent"
```

```xml
5      android:layout_height="match_parent"
6      android:orientation="vertical"
7      tools:context=".fragment.VideoFragment">
8      <ListView
9          android:id="@+id/listview_video"
10         android:layout_width="match_parent"
11         android:layout_height="wrap_content"/>
12 </LinearLayout>
```

【代码 4-4】 video_item_list.xml

```xml
1  <?xml version="1.0" encoding="utf-8"?>
2  <LinearLayout xmlns:android="http://schemas.android.com/apk/res/android"
3      android:layout_width="match_parent"
4      android:layout_height="wrap_content"
5      android:orientation="vertical">
6      <RelativeLayout
7          android:layout_width="match_parent"
8          android:layout_height="200dp">
9          <ImageView
10             android:id="@+id/video_item_pic"
11             android:layout_width="match_parent"
12             android:layout_height="wrap_content"
13             android:scaleType="fitXY"
14             android:src="@drawable/video1" />
15         <TextView
16             android:id="@+id/video_item_time"
17             android:layout_width="wrap_content"
18             android:layout_height="wrap_content"
19             android:layout_alignParentRight="true"
20             android:layout_alignParentBottom="true"
21             android:layout_marginRight="5dp"
22             android:text="4:10"
23             android:textColor="#FFFFFF"
24             android:textSize="20dp" />
25     </RelativeLayout>
26     <LinearLayout
27         android:layout_width="match_parent"
28         android:layout_height="0dp"
29         android:layout_weight="1"
30         android:orientation="horizontal">
31         <TextView
32             android:id="@+id/video_item_title"
33             android:layout_width="0dp"
34             android:layout_height="match_parent"
35             android:layout_weight="3"
```

```
36              android:gravity = "center_vertical"
37              android:textSize = "15sp" />
38          <Button
39              android:id = "@ + id/video_item_button"
40              android:layout_width = "0dp"
41              android:layout_height = "wrap_content"
42              android:layout_weight = "1"
43              android:text = "播放"
44              android:textSize = "8pt" />
45      </LinearLayout>
46  </LinearLayout>
```

②在 adapter 包中,创建 VideoAdapter 适配器用于视频列表项的加载。

【代码 4 – 5】 VideoAdapter. java

```
1   public class VideoAdapter extends BaseAdapter {
2       private Context context;
3       private List <Video > list;
4       LayoutInflater inflater;
5
6       public VideoAdapter(Context context,List <Video > list) {
7           this.context = context;
8           this.list = list;
9           inflater = LayoutInflater.from(context);
10      }
11      @Override
12      public int getCount() {
13          return list.size();
14      }
15      @Override
16      public Object getItem(int arg0) {
17          return list.get(arg0);
18      }
19      @Override
20      public long getItemId(int arg0) {
21          return arg0;
22      }
23      @Override
24      public View getView(final int position,View view,ViewGroup arg2) {
25          if (view = = null) {
26              view = inflater.inflate(R.layout.video_item_list,null);
27          }
28          ImageView video_item_pic = view.findViewById(R.id.video_item_pic);
29          TextView video_item_time = view.findViewById(R.id.video_item_time);
30          TextView video_item_title = view.findViewById(R.id.video_item_title);
31          Button video_item_button = view.findViewById(R.id.video_item_button);
```

```
32      video_item_pic.setImageResource(list.get(position).getPic());
33      int time = list.get(position).getTime();
34      //计算视频分秒数
35      int minute = time/60;
36      int second = time% 60;
37      if(second < 10){
38          video_item_time.setText(minute + ":" + "0" + second);
39      }else{
40          video_item_time.setText(minute + ":" + second);
41      }
42      video_item_title.setText(list.get(position).getTitle());
43      video_item_button.setOnClickListener(new View.OnClickListener() {
44          @Override
45          public void onClick(View view) {
46              //单击播放列表跳转到视频播放页面VideoActivity
47              Intent intent = new Intent();
48              intent.setAction("com.example.learntopass.VideoActivity");
49              intent.putExtra("uri",list.get(position).getUri());
50              context.startActivity(intent);
51          }
52      });
53      return view;
54  }
55 }
```

③在 VideoFragment.java 中，编写功能逻辑代码。

【代码4-6】 VideoFragment.java

```
1  public class VideoFragment extends Fragment {
2      private ListView listview_video;
3      private ArrayList<Video> list = new ArrayList<>();
4      @Override
5      public View onCreateView(LayoutInflater inflater,ViewGroup container,
6      Bundle savedInstanceState) {
7          return inflater.inflate(R.layout.fragment_video,container,false);
8      }
9      @Override
10     public void onViewCreated(View view,Bundle savedInstanceState) {
11         super.onViewCreated(view,savedInstanceState);
12         initData();
13         listview_video = view.findViewById(R.id.listview_video);
14         VideoAdapter adapter = new VideoAdapter(getContext(),list);
15         listview_video.setAdapter(adapter);
16
```

```
17      }
18      private void initData(){
19          Video video1 = new Video();
20          video1.setUri("android.resource://"
21                  +getActivity().getPackageName()+"/"+R.raw.video1);
22          video1.setTitle("泱泱大国 历史悠久 文件博大");
23          video1.setPic(R.drawable.video1);
24          video1.setTime(420);
25          list.add(video1);
26          //省略其他数据初始化代码
27      }
28  }
```

(3) 完成视频播放页面功能代码编写。

在 VideoAdapter.java 中的"播放"按钮单击事件中，跳转至 VideoActivity 视频播放页面，下面完成该页面的布局及功能逻辑代码编写。

①在 activity_video.xml 布局文件中，编写 VideoActivity 的布局。

【代码 4-7】 activity_video.xml

```
1   <?xml version="1.0" encoding="utf-8"?>
2   <RelativeLayout xmlns:android="http://schemas.android.com/apk/res/android"
3       android:orientation="vertical"
4       android:layout_width="match_parent"
5       android:layout_height="match_parent"
6       android:background="#000000" >
7       <VideoView
8           android:id="@+id/videoview"
9           android:layout_width="match_parent"
10          android:layout_height="wrap_content"
11          android:layout_centerInParent="true"/>
12  </RelativeLayout>
```

②在 VideoActivity.java 中，编写视频播放功能逻辑代码。

【代码 4-8】 VideoActivity.java

```
1   public class VideoActivity extends AppCompatActivity {
2       private VideoView videoView;
3       @Override
4       protected void onCreate(Bundle savedInstanceState) {
5           super.onCreate(savedInstanceState);
6           setContentView(R.layout.activity_video);
7           videoView = findViewById(R.id.videoview);
8           //获取视频路径地址
9           Intent intent = getIntent();
10          String uriString = intent.getStringExtra("uri");
11          Uri uri = Uri.parse(uriString);
```

```
12          videoView.setVideoURI(uri);
13          MediaController controller = new MediaController(this);
14          videoView.setMediaController(controller);
15          videoView.start();
16      }
17 }
```

(4) 完成音频列表功能代码编写。

①编写 AudioFragment 的布局及视频列表的列表项布局。

AudioFragment 的布局文件为"fragment_audio.xml"，布局中包含一个 ListView 控件，其代码与 fragment_video.xml 基本相同；创建列表项布局文件"audio_item_list.xml"，包含 2 个 TextView，分别用于显示音频标题和时长，具体代码省略。

②在 adapter 包中，创建 AudioAdapter 适配器，用于音频列表项的加载。

【代码4-9】 AudioAdapter.java

```
1  public class AudioAdapter extends BaseAdapter {
2      private List<Audio> audioList;
3      private Context context;
4      private LayoutInflater inflater;
5      private MediaPlayer mediaPlayer;
6      public AudioAdapter(Context context,List<Audio> audioList) {
7          this.context = context;
8          this.audioList = audioList;
9          inflater = LayoutInflater.from(context);
10     }
11     @Override
12     public int getCount() {
13         return audioList.size();
14     }
15     @Override
16     public Object getItem(int position) {
17         return audioList.get(position);
18     }
19     @Override
20     public long getItemId(int position) {
21         return position;
22     }
23
24     @Override
25     public View getView(final int position,View convertView,
26                         ViewGroup parent) {
27         View view = inflater.inflate(R.layout.audio_item_list,null);
28         TextView titleTextView = view.findViewById(R.id.tv_audio_title);
29         TextView timeTextView = view.findViewById(R.id.tv_audio_time);
30         titleTextView.setText(audioList.get(position).getTitle());
```

```
31        //获取当前音频时长并计算分钟数和秒数
32        int time = audioList.get(position).getStrTime();
33        int minute = time/60;
34        int second = time % 60;
35        if (second < 10) {
36            timeTextView.setText(minute + ":" + "0" + second);
37        } else {
38            timeTextView.setText(minute + ":" + second);
39        }
40        return view;
41    }
42 }
```

③在 AudioFragment.java 中，编写功能逻辑代码。

【代码 4 – 10】 AudioFragment.java

```
1  public class AudioFragment extends Fragment {
2      private  ArrayList<Audio> list = new ArrayList<>();
3      private ListView listView;
4      @Override
5      public void onCreate(Bundle savedInstanceState) {
6          super.onCreate(savedInstanceState);
7
8      }
9      @Override
10     public View onCreateView(LayoutInflater inflater,ViewGroup container,
11     Bundle savedInstanceState) {
12         return inflater.inflate(R.layout.fragment_audio,container,false);
13     }
14     @Override
15     public void onViewCreated(View view,Bundle savedInstanceState) {
16         super.onViewCreated(view,savedInstanceState);
17         initData();
18         listView = view.findViewById(R.id.listview_audio);
19         AudioAdapter adapter = new AudioAdapter(getContext(),list);
20         listView.setAdapter(adapter);
21         //添加列表项单击事件监听
22         listView.setOnItemClickListener(new AdapterView.OnItemClickListener() {
23             @Override
24             public void onItemClick(AdapterView<?> parent,View view,
25                                    int position,long id) {
26                 //采用隐式 Intent,将当前列表位置和音频数据传递给下个页面
27                 Intent intent = new Intent();
28                 intent.setAction(
29                         "com.example.learntopass.AudioActivity");
```

```
30                intent.putExtra("position",position);
31                intent.putExtra("data",list);
32                startActivity(intent);
33            }
34        });
35    }
36
37    private void initData(){
38        Audio audio1 = new Audio();
39        audio1.setRawId(R.raw.voice1);
40        audio1.setStrTime(200);
41        audio1.setTitle("以科技创新支撑引领高质量发展");
42                audio1.setContent("科技是国家强盛之基,创新是引领发展的" +
43                        "第一动力。以科技创新支撑引领高质量发展,是破解" +
44                        "当前经济社会发展深层次矛盾和问题的必然选择,也是" +
45                        "加快转变经济发展方式、调整社会结构、提高发展质量" +
46                        "和效益的重要抓手。进入新发展阶段,要贯彻新发展理" +
47                        "念,构建新发展格局,着力加强科技创新基础研究,着" +
48                        "力以改革激发创新创造活力,着力推动科技与经济深度" +
49                        "融合,促进科技实力大幅跃升,形成推动高质量发展的" +
50                        "有力支撑。");
51        list.add(audio1);
52        //省略其他数据
53    }
54 }
```

(5) 完成音频播放页面功能代码编写。

在 AudioAdapter.java 中的列表项单击事件中,跳转至 AudioActivity 视频播放页面,下面完成该页面的布局及功能逻辑代码编写。

①在 activity_audio.xml 布局文件中,编写 AudioActivity 的布局。

布局中包括 2 个 TextView,用于显示音频标题和内容,下方有 3 个 ImageView,用于显示"播放/暂停"按钮、"前一音频"和"后一音频"按钮。

【代码 4 – 11】 activity_audio.xml

```
1  <?xml version="1.0" encoding="utf-8"?>
2  <LinearLayout xmlns:android="http://schemas.android.com/apk/res/android"
3      xmlns:app="http://schemas.android.com/apk/res-auto"
4      xmlns:tools="http://schemas.android.com/tools"
5      android:layout_width="match_parent"
6      android:layout_height="match_parent"
7      tools:context=".activity.AudioActivity"
8      android:background="#000000"
```

```xml
9        android:padding = "20dp"
10       android:orientation = "vertical" >
11       <TextView
12           android:id = "@ + id/tv_title"
13           android:layout_width = "match_parent"
14           android:layout_height = "wrap_content"
15           android:gravity = "center"
16           android:text = "这是标题"
17           android:textColor = "#FFFFFF"
18           android:textSize = "25sp" />
19       <TextView
20           android:id = "@ + id/tv_content"
21           android:layout_width = "match_parent"
22           android:layout_height = "0dp"
23           android:layout_weight = "1"
24           android:layout_marginTop = "20dp"
25           android:text = "这是内容"
26           android:textColor = "#FFFFFF"
27           android:textSize = "20sp" />
28       <RelativeLayout
29           android:id = "@ + id/rv_control"
30           android:layout_width = "match_parent"
31           android:layout_height = "wrap_content" >
32           <ImageView
33               android:id = "@ + id/iv_play_pause"
34               android:layout_width = "80dp"
35               android:layout_height = "80dp"
36               android:src = "@drawable/audioplay"
37               android:layout_centerInParent = "true" />
38           <ImageView
39               android:id = "@ + id/iv_previous"
40               android:layout_width = "80dp"
41               android:layout_height = "80dp"
42               android:src = "@drawable/audioprevious" />
43           <ImageView
44               android:id = "@ + id/iv_next"
45               android:layout_width = "80dp"
46               android:layout_height = "80dp"
47               android:src = "@drawable/audionext"
48               android:layout_alignParentRight = "true" />
49       </RelativeLayout>
50   </LinearLayout>
```

②在 AudioActivity.java 中，编写音频播放功能逻辑代码。

【代码 4-12】 AudioActivity.java

```java
1   public class AudioActivity extends AppCompatActivity {
2
3       private int currentPosition;
4       private ArrayList<Audio> list;
5       private ImageView audioPlay,audioNext,audioPrevious;
6       private TextView audioTitle,audioContent;
7       private MediaPlayer player = null;
8
9       @Override
10      protected void onCreate(Bundle savedInstanceState) {
11          super.onCreate(savedInstanceState);
12          setContentView(R.layout.activity_audio);
13          initView();
14          //获取当前音频position及音频数据集合
15          Intent intent = getIntent();
16          currentPosition = intent.getIntExtra("position",0);
17          list = (ArrayList<Audio>) intent.getSerializableExtra("data");
18
19          audioTitle.setText(list.get(currentPosition).getTitle());
20          audioContent.setText(list.get(currentPosition).getContent());
21
22          startAudio(this,list.get(currentPosition).getRawId());
23          audioPlay.setImageResource(R.drawable.audiopause);
24      }
25      //初始化控件并设置事件监听
26      void initView() {
27          audioPlay = findViewById(R.id.iv_play_pause);
28          audioNext = findViewById(R.id.iv_next);
29          audioPrevious = findViewById(R.id.iv_previous);
30          audioTitle = findViewById(R.id.tv_title);
31          audioContent = findViewById(R.id.tv_content);
32          //播放/暂停按钮事件监听
33          audioPlay.setOnClickListener(new View.OnClickListener() {
34              @Override
35              public void onClick(View v) {
36                  //根据返回结果更改按钮图片,true为暂停,false为播放
37                  boolean isPause = pauseAudio();
38                  if(isPause){
39                      audioPlay.setImageResource(R.drawable.audioplay);
40                  }else{
41                      audioPlay.setImageResource(R.drawable.audiopause);
42                  }
43              }
44          });
45          //下一个音频按钮事件监听
46          audioNext.setOnClickListener(new View.OnClickListener() {
47              @Override
48              public void onClick(View v) {
```

```
49              stopAudio();//停止原音频播放
50              currentPosition =(currentPosition +1)% list.size();
51              audioTitle.setText(list.get(currentPosition).getTitle());
52              audioContent.setText(list.get(currentPosition).
53                                    getContent());
54              startAudio(list.get(currentPosition).getRawId());
55          }
56        });
57        //前一个音频按钮事件监听
58        audioPrevious.setOnClickListener(new View.OnClickListener() {
59          @Override
60          public void onClick(View view) {
61              stopAudio();
62              currentPosition = currentPosition -1;
63              if(currentPosition = = -1){
64                  currentPosition = list.size() -1;
65              }
66              audioTitle.setText(list.get(currentPosition).getTitle());
67              audioContent.setText(list.get(currentPosition).
68                                    getContent());
69              startAudio(list.get(currentPosition).getRawId());
70          }
71        });
72    }
73
74    @Override
75    protected void onDestroy() {
76        super.onDestroy();
77        stopAudio();
78    }
79    //创建并启动音频
80    public void startAudio(Context context,int audioId){
81      if(player = =null) {
82          player = MediaPlayer.create(context,audioId);
83      }
84      player.start();
85    }
86
87    //暂停及播放音频状态切换
88    public boolean pauseAudio(){
89        if (player! =null && player.isPlaying()) {
90            player.pause();
91            return true;     //暂停状态
92        }else if(player! =null && (! player.isPlaying())){
93            player.start();
94            return false;    //播放状态
95        }
96      return false;
```

```
97          }
98      //停止并释放音频资源
99      public void stopAudio(){
100         if(player! =null){
101             player.stop();
102             player.release();
103             player=null;
104         }
105     }
106 }
107
108
109
```

任务反思

编写并运行程序,将在代码编写及程序调试过程中出现的异常信息、产生原因及解决方法记录在下方。

问题1:_____

产生原因:_____

_____

解决方法:_____

_____

问题2:_____

产生原因:_____

_____

解决方法:_____

_____

任务总结及巩固

小白:师兄,VideoView 和 MediaPlayer 的使用看起来并不难,但是我在使用 MediaPlayer 的时候还真遇到了问题,经常出现非法状态的问题。

师兄:我猜想你也会问我这个问题的。MediaPlayer 看起来并不难,但是在使用的时候要格外注意,它的状态变化是有特定要求的,并不是可以随心所欲地进行播放、暂停方法的调用。但是只要按照图 4-8 所示进行播放、暂停等状态的改变,就不会出错。

功能模块 4　电台模块的实现

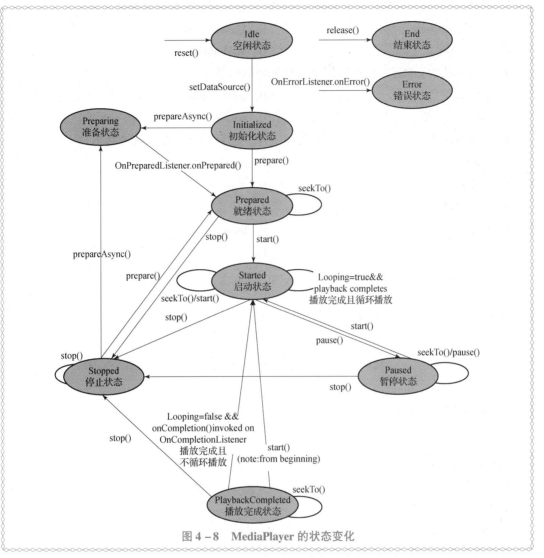

图 4-8　MediaPlayer 的状态变化

小白：原来奥秘在这里呢！怪不得我会出错，在使用的时候，确实有时没有按照状态调用方法。

师兄：世间万物有其规则，每个控件也都有它的使用规则，要掌握好规则，才能正确地使用。其实，MediaPlayer 也能够播放视频。但是需要与 SurfaceView 配合使用才行，这次给你的自学任务就是使用 SurfaceView + MediaPlayer 实现视频播放。

一、基础巩固

1. 下列方法中，（　　）可以用来为 VideoView 设置视频路径。

A. setVideoPath( )　　　　　　　　B. create( )

C. setDataSource( )　　　　　　　D. setPath( )

2. 下面（　　）控件可以与 VideoView 配合使用添加视频播放控制功能。

   A. MediaController　　　　　　　　B. VideoController

   C. Controller　　　　　　　　　　　D. MediaManager

3. 下列方法中，（　　）可以用来获取视频的时长。

   A. getTime( )　　　　　　　　　　　B. getDuration( )

   C. isPlaying( )　　　　　　　　　　D. pause ( )

4. 下列方法中，（　　）可以解析音频文件，使音频播放进入就绪状态。

   A. prepare( )　　　　　　　　　　　B. pause( )

   C. seekTo( )　　　　　　　　　　　D. start( )

5. 下列方法中，（　　）可以定位到音频的某个位置。

   A. prepare( )　　　　　　　　　　　B. pause( )

   C. seekTo( )　　　　　　　　　　　D. start( )

二、技术实践

使用 SurfaceView 和 MediaPlayer 实现视频播放。

任务 2　服务及广播的应用

任务描述

通过服务与广播相关技术的学习，实现通过服务实现音频的播放功能。

任务学习目标

通过本任务需达到以下目标：

➢ 能够使用 Service 来实现音频播放。

➢ 能够了解 Service 生命周期对应方法的基本用法。

➢ 能够掌握 BroadcastReceiver 广播运行机制。

技术储备

一、Service

（一）认识服务

在前面实现的音频播放功能中，当我们退出程序界面，音频播放就停止了，但在使用音乐播放器时，即便程序界面退出，音乐依然不绝于耳；当在程序中下载文件时，程序退出，下载也依然继续。这些功能都需要使用 Service（服务）来实现。

Service 是 Android 四大组件之一，是能够长期运行在后台的、没有用户界面的应用组件。Service 一般由 Activity 启动，但并不依赖于 Activity。当 Activity 的生命周期结束，Service 仍然会正常运行，直到自己的生命周期结束。因此，Service 适合执行一段时间内不需要显示界面的后台耗时操作，比如下载网络数据、播放音乐等。

Service 运行在主线程中，并不会自动开启线程。如果在 Service 中需要处理耗时的操作，则需要在服务的内部启动子线程进行处理，否则就会出现主线程被阻塞的情况。

（二）创建服务

在项目中创建 Service 通常需要"创建"和"注册"两个步骤。当然，在 Android Studio 中，也可以通过右击程序包名，依次选择 New→Service→Service 进行创建，将两个步骤合二为一。

下面我们看一下创建好的 Service。

```
1   public class MyService extends Service {
2       public MyService() {
3       }
4
5       @Override
6       public IBinder onBind(Intent intent) {
7           //TODO: Return the communication channel to the service.
8           throw new UnsupportedOperationException("Not yet implemented");
9       }
10  }
```

从上述代码中可以看到，MyService 继承自 Service 类，并重写了 onBind( )方法。OnBind( )方法是创建 Service 时必须要实现的。

与 Activity 类似，Service 创建之后也需要在 AndroidManifest.xml 中进行注册，具体代码如下。

```
1   <application
2           ......>
3       ......
4       <service
5           android:name = "com.example.service.MyService"
6           android:enabled = "true"
7           android:exported = "true" >
8       </service>
9   </application>
```

上述代码中，<service></service>标签中设置了 3 个属性："android：name"为 Service 的路径，"android：enabled"表示系统是否能够实例化该服务，"android：exported"表示该服务是否能够被其他应用程序中的组件调用或进行交互。

（三）启动服务

创建服务后还需要启动服务才能够开始后台运行。服务的启动方式有两种，分别是调用 startService( )方法启动服务和调用 bindService( )方法启动服务。

1. 调用 startService( )方法启动服务

采用 startService( )方法启动服务，需要在前面创建的 Service 类中重写 onStartCommand( )方法，并在 Activity 中调用 startService( )方法启动服务，调用 stopService( )方法停止服务。startService( )方法和 stopService( )方法都接收 Intent 作为参数。

下面通过一个利用 Service 生成两个随机数并求和的案例，实现上述过程。

(1) 在 MyService 类中，重写 onCreate( )、onStartCommand( )、onDestroy( )方法。

将生成随机数并求和的语句写在 onStartCommand( )方法中，并在 3 个方法中分别加入 Log 语句打印日志。

```
1   public class MyService extends Service {
2       public MyService() {
3       }
4       @Override
5       public void onCreate() {
6           super.onCreate();
7           Log.e("后台服务","onCreate 方法执行了");
8       }
9       @Override
10      public int onStartCommand(Intent intent,int flags,int startId) {
11          Log.e("后台服务","onStartCommand 方法执行了");
12          double random1 = Math.random()*100;
13          double random2 = Math.random()*100;
14          double sum = random1 + random2;
15          Log.e("后台服务","随机数求和结果" + sum);
16          return super.onStartCommand(intent,flags,startId);
17      }
18      @Override
19      public void onDestroy() {
20          super.onDestroy();
21          Log.e("后台服务","onDestroy 方法执行了");
22
23      }
24      @Override
25      public IBinder onBind(Intent intent) {
26          //TODO: Return the communication channel to the service.
27          throw new UnsupportedOperationException("Not yet implemented");
28      }
29  }
```

(2) 在布局中加入两个按钮，用于启动服务和停止服务。并在 MainActivity.java 中，为按钮添加单击事件监听。

```
1   public class MainActivity extends AppCompatActivity {
2       @Override
3       protected void onCreate(Bundle savedInstanceState) {
4           super.onCreate(savedInstanceState);
5           setContentView(R.layout.activity_main);
6           ButtonstartButton = findViewById(R.id.bt_start);
7           ButtonstopButton = findViewById(R.id.bt_stop);
8           startButton.setOnClickListener(new View.OnClickListener() {
9               @Override
```

```
10          public void onClick(View view) {
11              Intent intent = new Intent(MainActivity.this,
12                          MyService.class);
13              startService(intent);
14          }
15      });
16      stopButton.setOnClickListener(new View.OnClickListener() {
17          @Override
18          public void onClick(View view) {
19              Intent intent = new Intent(MainActivity.this,
20                          MyService.class);
21              stopService(intent);
22          }
23      });
```

上述代码中，第 11~13 行，在单击"启动"按钮时，创建 Intent 对象，调用 startService( ) 方法启动服务；第 19~21 行，在单击"停止"按钮时，同样创建 Intent 对象，调用 stopService( ) 方法停止服务。

运行程序，单击"启动"按钮，此时在 LogCat 中打印信息如图 4-9 所示。可以看到，在启动服务时，onCreate( ) 和 onStartCommand( ) 方法依次执行。

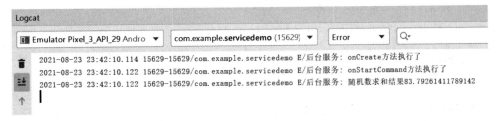

图 4-9 单击"启动"按钮日志打印情况

当单击"停止"按钮时，onDestroy( ) 方法执行，服务被销毁。

在采用这种方式启动服务时，启动服务后，Activity 与 Service 就没有关联了，服务会一直处于运行状态，Activity 无法控制 Service 的执行。如果希望在 Activity 中指挥 Service 的执行，就需要使用第 2 种方法。

2. 调用 bindService( ) 方法启动服务

采用 bindService( ) 方法启动服务会使 Activity 与 Service 绑定，在 Activity 中可以通过 Service 对象调用 Service 中的方法控制服务的执行。采用这种方式就需要借助 Service 类中的 onBind( ) 方法。

下面对生成随机数的案例进行改写，服务提供两个功能：生成随机数和求和，通过 Activity 先后调用两个方法，并将结果显示在界面上。

（1）在 Service 类中，创建 Binder 对象，对服务的功能进行管理。

在 Service 类中，创建 MyBinder 继承自 Binder，将服务的两个功能在 MyBinder 中实现。

```
1   public class MyService extends Service{
2
3       class MyBinder extends Binder{
4           //生成随机数
5           double getRadom(){
6               return Math.random()*100;
7           }
8           //随机数求和
9           double sum(double r1,double r2){
10              return r1+r2;
11          }
12      }
13      //绑定服务
14      @Override
15      public IBinder onBind(Intent intent){
16          Log.i("后台服务","onBind");
17          return new MyBinder();
18      }
19      //解绑服务
20      @Override
21      public boolean onUnbind(Intent intent){
22          Log.i("后台服务","onUnbind");
23          return super.onUnbind(intent);
24      }
25      //省略其他方法
26  }
```

上述代码中，第3~12行创建了MyBinder类，其继承自Binder，在类中定义了getRadom()方法和sum()方法，分别用于生成随机数和求和；第15~18行的onBind()方法的返回值类型为IBinder，IBinder是一个接口，Binder是IBinder的实现类，自然MyBinder也实现了IBinder接口，在onBind()方法中创建MyBinder对象，返回。

第21~24行重写了onUnbind()方法，在服务解绑时调用。

（2）在Activity中，调用bindService()方法绑定服务。

bindService()方法的具体定义如下所示，需要3个参数。

`boolean bindService(Intent service,ServiceConnection conn,int flags)`

● Intent service：用于指定要启动的Service的Intent对象；

● ServiceConnection conn：用于监听Activity与Service之间的连接状态；当Activity与Service连接成功时，调用conn的onServiceConnected()方法，该方法的第2个参数即为onBind()方法中返回的IBinder对象；当Activity与Service断开连接时，调用conn的onServiceDisconnected()方法。

● int flags：表示Activity绑定服务时是否自动创建Service对象。该参数为常量"BIND_AUTO_CREATE"时表示自动创建，参数为0时表示不自动创建。

在 Activity 中，将按照如下步骤进行服务绑定。

①创建 ServiceConnection 对象，实现两个方法。

②通过 IBinder 对象调用 Service 中的方法，实现对服务的控制。

具体代码如下所示。

```
1   public class MainActivity extends AppCompatActivity {
2       private MyService.MyBinder myBinder;
3       private Button bindButton、radomButton、unbindButton;
4       private TextView radomTextView1;
5       private TextView radomTextView2;
6       private TextView sumTextView;
7       private ServiceConnection conn = new ServiceConnection() {
8           @Override
9           public void onServiceConnected(ComponentName name,IBinder service){
10              myBinder = (MyService.MyBinder)service;
11          }
12          @Override
13          public void onServiceDisconnected(ComponentName name) {
14          }
15      });
16      @Override
17      protected void onCreate(Bundle savedInstanceState) {
18          super.onCreate(savedInstanceState);
19          setContentView(R.layout.activity_main);
20          bindButton = findViewById(R.id.button);     //绑定服务
21          radomButton = findViewById(R.id.button2);   //生成随机数
22          unbindButton = findViewById(R.id.button3);//解除绑定
23          radomTextView1 = findViewById(R.id.tv_radom1);//显示第一个随机数
24          radomTextView2 = findViewById(R.id.tv_radom2);//显示第二个随机数
25          sumTextView = findViewById(R.id.tv_sum);    //显示随机数之和
26          bindButton.setOnClickListener(new View.OnClickListener() {
27              @Override
28              public void onClick(View v) {
29                  //绑定服务
30                  Intent intent = new Intent(MainActivity.this
31                                              ,RadomService.class);
32                  bindService(intent,conn,BIND_AUTO_CREATE);
33              }
34          });
35          radomButton.setOnClickListener(new View.OnClickListener() {
36              @Override
37              public void onClick(View v) {
38                  double radom1 =  myBinder.getRadom();//生成第一个随机数
39                  double radom2 =myBinder.getRadom();//生成第二个随机数
```

```
40              double sum = myBinder.sum(radom1,radom2);   //求和
41              radomTextView1.setText("第一个随机数为" + radom1);
42              radomTextView2.setText("第二个随机数为" + radom2);
43              sumTextView.setText("随机数之和为" + sum);
44           }
45        });
46        unbindButton.setOnClickListener(new View.OnClickListener() {
47           @Override
48           public void onClick(View v) {
49              //取消绑定服务
50              Intent intent = new Intent(MainActivity.this,
51                                    RadomService.class);
52              unbindService(conn);
53           }
54        });
55    }
56 }
```

上述代码中，第7~16行代码创建了ServiceConnection对象，重写了onServiceConnected()和onServiceDisconnected()方法，在onServiceConnected()中，将通过参数传递过来的IBinder接口实现类对象service转换为MyService类中定义的MyBinder对象。

第26~34行代码在绑定按钮单击事件中，调用bindService()方法绑定服务；第35~45行代码在按钮单击事件中，通过前面获取到的MyBinder对象，调用服务中的方法生成随机数并求和，实现了Activity与Service的通信；第47~55行代码中调用unbindService()方法解绑服务。

运行程序，单击绑定服务按钮，在Logcat中打印的日志信息如图4-10所示。在绑定服务时，onCreate()和onBind()方法依次被调用。

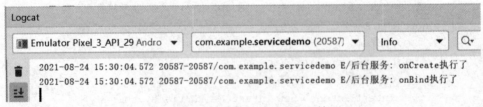

图4-10 单击绑定按钮日志打印情况

继续单击生成随机数按钮，界面中显示如图4-11所示的运算结果。调用服务中的相关方法并将结果在界面上更新。

单击解除绑定按钮，在Logcat中打印的日志信息如图4-12所示。在解除绑定服务时，onUnbind()和onDestroy()方法依次被调用，解绑并销毁服务。

图4-11 单击生成随机数显示效果

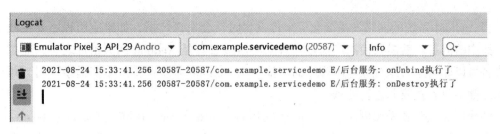

图 4-12　单击解除绑定按钮日志打印情况

二、BroadcastReceiver 广播

广播在生活中屡见不鲜。在英语考试时，通常学校广播站会播放听力音频，学生将收音机调至特定频道就能够接收到音频信号。通过广播能够方便、快捷地将消息传递给接收者。Android 系统中也引入了广播机制，用于在组件之间传递消息。

（一）认识广播机制

在 Android 系统中，当电池电量低、耳机插拔、关闭或打开飞行模式时都会发送一条广播，这些是系统广播；应用程序也可以自己发送广播，这些是普通广播。Android 提供了一整套 API，允许应用程序自由地发送和接收广播。Android 中的每个应用程序都可以对自己感兴趣的广播进行注册，这样，该程序就只会接收自己所关心的广播内容。

Android 中的广播使用了设计模式中的观察者模式，这是一种基于消息的发布／订阅事件模型，模型中有 3 个角色：消息发布者（广播发送者）、消息订阅者（广播接收者）和处理中心（Activity Manager Service，AMS）。广播机制的具体实现流程如图 4-13 所示。

图 4-13　广播机制示意图

广播接收者在 AMS 中进行注册，广播发送者向 AMS 发送广播。AMS 查找符合相应条件的广播接收者（BroadcastReceiver），将广播发送到相应的消息循环队列中；广播接收者通过消息循环接收到此广播，并进行处理。在此过程中，广播发送者和广播接收者是异步的，广播发送者不需关心有无接收者，也不需确定接收者是否收到广播。

在同一个 App 具有多个进程的不同组件之间进行消息通信、在不同 App 的组件之间进行消息通信以及 Android 系统在特定情况下与 App 之间进行消息通信的场景，都非常适合使用广播机制进行处理。

（二）广播接收者

当 Android 系统的网络发生变化等事件发生时，会发送广播，如果应用程序要接收并过滤广播中的消息，就需要使用广播接收者。广播接收者是 Android 四大组件之一，通过广播接收者可以监听来自系统或应用程序的广播。当 Android 系统产生一个广播事件时，可以有多个广播接收者接收并进行处理。

创建广播接收者需要创建一个类继承自 BroadcastReceiver，并重写父类的 onReceive( ) 方法。当有广播到来时，onReceive( ) 方法就会得到执行，接收广播后需要实现的功能逻辑可以在这个方法中处理。除此之外，广播接收者还需要对自己感兴趣的广播进行注册，注册广播一般有两种方式：在代码中动态注册和在 AndroidManifest.xml 文件中静态注册。

1. 动态注册

下面使用动态注册的方式编写一个能够监听网络变化的程序。

（1）创建广播接收者。

创建 NetworkBroadcastReceiver 类继承 BroadcastReceiver，并重写 onReceive( ) 方法。在方法中，实现接收到广播后，弹出提示信息。

```
1   public class NetworkBroadcastReceiver extends BroadcastReceiver{
2       @Override
3       public void onReceive(Context context,Intent intent){
4           Toast.makeText(context,"网络状态改变",Toast.LENGTH_SHORT).show();
5           ConnectivityManager manager =(ConnectivityManager)
6                   context.getSystemService(Context.CONNECTIVITY_SERVICE);
7           NetworkCapabilities cap =manager.getNetworkCapabilities(
8                                       manager.getActiveNetwork());
9           if(cap!=null &&
10              cap.hasTransport(NetworkCapabilities.TRANSPORT_CELLULAR)){
11              Toast.makeText(context,"移动网络可用",
12                                      Toast.LENGTH_SHORT).show();
13          }else{
14              Toast.makeText(context,"移动网络不可用",
15                                      Toast.LENGTH_SHORT).show();
16          }
17      }
18  }
```

在 onReceive( ) 方法中，获取 ConnectivityManager 对象及 NetworkCapabilities 对象进行网络状态判断，第 10 行中的 hasTransport( ) 方法用于判断网络是否可用，其值为常量"NetworkCapabilities.TRANSPORT_CELLULAR"时代表移动网络。

（2）进行动态注册。

在 MainActivity.java 中，调用 registerReceiver( ) 方法进行广播的动态注册。

```
1   public class MainActivity extends AppCompatActivity {
2       private NetworkBroadcastReceiver receiver;
3       @Override
4       protected void onCreate(Bundle savedInstanceState) {
5           super.onCreate(savedInstanceState);
6           setContentView(R.layout.activity_main);
7           receiver = newNetworkBroadcastReceiver();
8
9           //动态注册
10          IntentFilter filter = new IntentFilter();
11          filter.addAction("android.net.conn.CONNECTIVITY_CHANGE");
12          registerReceiver(receiver, filter);
13      }
14      @Override
15      protected void onDestroy() {
16          super.onDestroy();
17          unregisterReceiver(receiver);
18      }
19  }
```

在上述代码中，第 10~11 行创建了 IntentFilter 对象，并调用 addAction() 方法设置 action，这里 action 的值就是网络状态发生变化时，系统发出的广播常量 "android.net.conn.CONNECTIVITY_CHANGE"。广播接收者想要监听什么广播，就可以通过这种方法进行添加。第 12 行调用 registerReceiver() 方法将 NetworkBroadcastReceiver 对象和 IntentFilter 对象传进去，进行动态注册。这样，NetworkBroadcastReceiver 就能接收到所有值为 "android.net.conn.CONNECTIVITY_CHANGE" 的广播，实现网络状态变化的监听。

需要注意的是，动态注册的广播需要在 onDestroy() 方法中调用 unregisterReceiver() 方法取消注册。同时，动态注册的广播接收者是否被注销依赖于注册广播的组件，比如，在 Activity 中注册的广播接收者，当 Activity 被销毁时，广播接收者也随之被注销。

另外，查询系统的网络状态需要声明权限，需要在 AndroidManifest.xml 文件中汇总，加入如下权限声明。

```
<uses-permission android:name="android.permission.ACCESS_NETWORK_STATE"/>
```

采用动态注册的方式时，程序必须处于运行状态，这样才能监听并接收广播。运行程序，按下 Home 键，打开模拟器的 System settings→Network & internet→Mobile network，关闭 Mobile data 会弹出"移动网络不可用"的提示信息，重新打开会弹出"移动网络可用"的提示信息，如图 4-14 所示。

> **注意**：一般情况下，onReceive() 方法会涉及与其他组件之间的交互，如创建一条状态栏通知（Notification）、启动 Service 等。默认情况下，广播接收者运行在 UI 线程中，不允许开启线程，onReceive() 方法不能执行耗时操作。

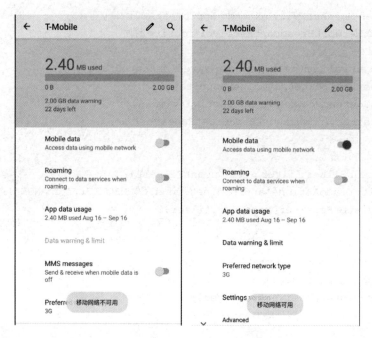

图 4-14 网络状态改变

2. 静态注册

采用静态注册广播接收者时,同样需要先创建广播接收者,具体方法与动态注册中相同。下面代码创建了一个广播接收者,用于监听开机系统广播。

```java
1  public class BootCompleteReceiver extends BroadcastReceiver{
2      @Override
3      public void onReceive(Context context,Intent intent){
4          Toast.makeText(context,"开机启动了",Toast.LENGTH_LONG).show();
5      }
6  }
```

注册时,在 AndroidManifest.xml 文件中注册广播接收者,下面的代码对前面创建的 MyBroadcastReceiver 进行静态注册。

```xml
1  <manifest xmlns:android = "http://schemas.android.com/apk/res/android"
2      package = "com.example.broadcastdemo">
3      <application
4          ……>
5          <receiver
6              android:name = ".BootCompleteReceiver"
7              android:enabled = "true"
8              android:exported = "true">
9              <intent-filter>
10                 <action android:name = "android.intent.action.BOOT_COMPLETED"/>
11             </intent-filter>
```

```
12          </receiver>
13          ......
14      </application>
15      <uses-permission
16          android:name="android.permission.RECEIVE_BOOT_COMPLETED"/>
17  </manifest>
```

在上述代码中，第 5~12 行对 MyBroadcastReceiver 进行注册，其中 android：name 属性值为 MyBroadcastReceiver 的路径；在 Intent 过滤器中，通过 action 加入了想要接收的广播，"android. intent. action. BOOT_COMPLETED" 即为开机广播。监听系统开机广播也需要声明权限，在 15~16 行加入了权限声明。

采用静态注册方式时，无论应用程序是否处于运行状态，广播接收者都会对程序进行监听。但在 Android 8.0 之后，大部分广播都不能使用静态注册，因此，优先使用动态注册。

（三）发送自定义广播

除了系统广播，应用程序也可以发送自定义广播。Android 中的广播主要有两种类型：无序广播和有序广播。

无序广播是一种完全异步执行的广播，在广播发出之后，所有的广播接收者几乎都会在同一时刻接收到这条广播消息，它们之间没有任何先后顺序可言。无序广播效率比较高，但无法被截断。

有序广播是一种同步执行的广播，在广播发出之后，同一时刻只会有一个广播接收者能够收到这条消息，当这个广播接收者中的逻辑执行完毕，广播才会继续传递。此时的广播接收者是有先后顺序的，优先级高的广播接收者可以先收到广播消息，并且前面的广播接收者可以截断正在传递的广播，这时，后面的广播接收者就无法收到广播消息了。

发送广播需要借助 Intent，也就是广播发送者将广播的意图（Intent）通过 sendBroadcast() 或 sendOrderedBroadcast() 方法发送出去。下面我们分别来看两种广播的发送方法。

1. 发送无序广播

下面通过单击按钮发送广播，广播接收者接收广播后弹出提示信息的案例，实现无序广播的发送。

在发送广播之前，需要先定义一个广播接收者，准备接收广播。创建 MyBroadcastReceiver 继承 BroadcastReceiver，具体代码如下。

```
1  public class MyBroadcastReceiver extends BroadcastReceiver {
2      @Override
3      public void onReceive(Context context,Intent intent) {
4          Toast.makeText(context,"接收无序广播",Toast.LENGTH_LONG).show();
5      }
6  }
```

在 MainActivity. java 中，动态注册 MyBroadcastReceiver，代码如下。

```
1  public class MainActivity extends AppCompatActivity {
2      private MyBroadcastReceiver receiver;
3      @Override
4      protected void onCreate(Bundle savedInstanceState) {
5          super.onCreate(savedInstanceState);
6          setContentView(R.layout.activity_main);
7          receiver = new MyBroadcastReceiver();
8          //动态注册广播
9          IntentFilter filter = new IntentFilter();
10         filter.addAction("com.example.mybroadcastdemo.MY_BROADCAST");
11         registerReceiver(receiver,filter);
12     }
13     @Override
14     protected void onDestroy() {
15         super.onDestroy();
16         unregisterReceiver(receiver);
17     }
18 }
```

在上述代码中,第 10 行向 IntentFilter 中添加了一条值为 "com.example.mybroadcastdemo.MY_BROADCAST" 的广播,接下来在单击按钮时就发出这样一条广播。

在按钮被单击时,创建 Intent 对象,并调用 sendBroadcast( )方法发送广播,代码如下。

```
1  public class MainActivity extends AppCompatActivity {
2      private MyBroadcastReceiver receiver;
3      private Button button;
4      @Override
5      protected void onCreate(Bundle savedInstanceState) {
6          super.onCreate(savedInstanceState);
7          setContentView(R.layout.activity_main);
8          receiver = new MyBroadcastReceiver();
9          //动态注册广播
10         IntentFilter filter = new IntentFilter();
11         filter.addAction("com.example.mybroadcastdemo.MY_BROADCAST");
12         registerReceiver(receiver,filter);
13
14         button = findViewById(R.id.button);
15         button.setOnClickListener(new View.OnClickListener() {
16             @Override
17             public void onClick(View v) {
18                 Intent intent = new Intent(
19                     "com.example.mybroadcastdemo.MY_BROADCAST");
20                 sendBroadcast(intent);
21             }
22         });
```

```
23      }
24      @Override
25      protected void onDestroy() {
26          super.onDestroy();
27          unregisterReceiver(receiver);
28      }
29  }
```

在上述代码中，第 18~19 行创建了 Intent 对象，并将要发送的广播的值传入构造方法，这里传入的值一定要与前面加入 IntentFilter 中的值一致；第 20 行调用 sendBroadcast() 方法，将广播发送出去，这样所有监听"com.example.mybroadcastdemo.MY_BROADCAST"广播的广播接收者就会收到广播消息。

运行程序，单击按钮，就会弹出提示信息，如图 4-15 所示。

2. 发送有序广播

发送有序广播时，优先级高的广播接收者会先收到广播消息。下面再创建一个广播接收者 AnotherBroadcastReceiver，具体代码如下：

图 4-15　发送无序广播

```
1   public class AnotherBroadcastReceiver extends BroadcastReceiver {
2       @Override
3       public void onReceive(Context context,Intent intent) {
4           Toast.makeText(context,"接收有序广播",Toast.LENGTH_LONG).show();
5       }
6   }
```

在动态注册时，设置两个广播接收者的优先级。在发送广播时，调用 sendOrderedBroadcast() 方法。

```
1   public class MainActivity extends AppCompatActivity {
2       private MyBroadcastReceiver receiver;
3       private AnotherBroadcastReceiver anotherReceiver;
4       private Button button;
5       @Override
6       protected void onCreate(Bundle savedInstanceState) {
7           super.onCreate(savedInstanceState);
8           setContentView(R.layout.activity_main);
9           receiver = new MyBroadcastReceiver();
10          anotherReceiver = new AnotherBroadcastReceiver();
11          //动态注册广播
```

```
12        IntentFilter filter = new IntentFilter();
13        filter.addAction("com.example.mybroadcastdemo.MY_BROADCAST");
14        filter.setPriority(10);
15        registerReceiver(receiver,filter);
16
17        IntentFilter anotherfilter = new IntentFilter();
18        anotherfilter.addAction(
19                "com.example.mybroadcastdemo.MY_BROADCAST");
20        anotherfilter.setPriority(20);
21        registerReceiver(anotherReceiver,anotherfilter);
22
23        button = findViewById(R.id.button);
24        button.setOnClickListener(new View.OnClickListener() {
25            @Override
26            public void onClick(View v) {
27                Intent intent = new Intent(
28                    "com.example.mybroadcastdemo.MY_BROADCAST");
29                sendOrderedBroadcast(intent,null);
30            }
31        });
32    }
33    @Override
34    protected void onDestroy() {
35        super.onDestroy();
36        unregisterReceiver(receiver);
37        unregisterReceiver(anotherReceiver);
38    }
39 }
```

在上述代码中，第14行调用IntentFilter的setPriority()方法，为广播接收者MyBroadcastReceiver设置优先级为10，其属性值越大，优先级越高；第17～21行动态注册AnotherBroadcastReceiver广播接收者，其优先级设置为20，优先级高于MyBroadcastReceiver；第29行调用sendOrderedBroadcast()方法发送有序广播。

运行程序，单击按钮，我们可以看到如图4-16所示的运行结果。先弹出"接收有序广播"的提示信息，说明优先级高的AnotherBroadcastReceiver先接收到广播消息，AnotherBroadcastReceiver完成广播处理逻辑后，MyBroadcastReceiver再收到广播，弹出"接收无序广播"的提示信息。

在接收到有序广播时，广播接收者还可以调用abortBroadcast()方法截断广播，后面的广播接收者就无法再接收广播消息。

在AnotherBroadcastReceiver中，完成处理逻辑后调用abortBroadcast()方法终止广播传递，代码如下：

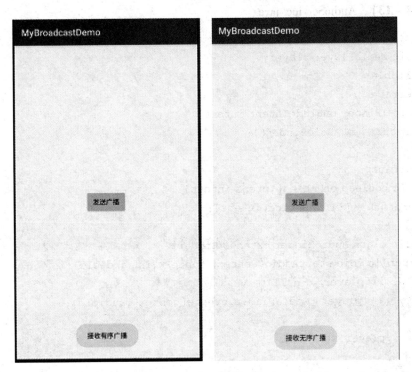

图 4-16 发送有序广播

```
1  public classAnotherBroadcastReceiver extends BroadcastReceiver {
2      @Override
3      public void onReceive(Context context,Intent intent) {
4          Toast.makeText(context,"接收有序广播",Toast.LENGTH_LONG).show();
5          abortBroadcast();
6      }
7  }
```

再次运行程序，只有 AnotherBroadcastReceiver 中的 Toast 信息能够弹出，这说明广播确实被终止传递了。

任务实施

1. 任务分析

本任务需要在项目中使用 Service 服务控制音频的播放，来实现音频文件在后台播放的功能。

2. 实现步骤

我们在任务 1 项目的基础上，继续实现本任务的相关功能。

（1）创建新包 service，在包中创建后台音频播放类 AudioService。

需要将原 AudioActivity.java 中对 MediaPlayer 的操作方法在 AudioService.java 中实现，并通过绑定服务对音频播放进行控制，具体代码如【代码 4-13】所示。

【代码 4-13】 AudioService.java

```java
public class AudioService extends Service {
    public MediaPlayer player;
    @Nullable
    @Override
    public IBinder onBind(Intent intent) {
        return new AudioBinder();
    }
    @Override
    public boolean onUnbind(Intent intent) {
        return super.onUnbind(intent);
    }
    public class AudioBinder extends Binder{
        public void startAudio(Context context,int audioId){
            if(player == null) {
                player = MediaPlayer.create(context,audioId);
            }
            player.start();
        }
        public boolean pauseAudio(){
            if (player! = null && player.isPlaying()) {
                player.pause();
                return true;    //暂停状态
            }else if(player! = null && (! player.isPlaying())){
                player.start();
                return false;    //播放状态
            }
            return false;
        }
        public void stopAudio(){
            if(player ! = null){
                player.stop();
                player.release();
                player = null;
            }
        }
    }
}
```

(2) 修改 AudioActivity.java 中的代码，绑定服务，获取 IBinder 对象，控制后台服务中的音频播放。

【代码 4-14】 AudioActivity.java

```java
1   public class AudioActivity1 extends AppCompatActivity {
2       private int currentPosition;
3       private ArrayList <Audio> list;
4       private ImageView audioPlay,audioNext,audioPrevious;
5       private TextView audioTitle,audioContent;
6       private MediaPlayer player = null;
7       private AudioService.AudioBinder audioService;   //IBinder 对象
8       private Intent serviceIntent;        //绑定服务所需 Intent 对象
9
10      //创建 ServiceConnection 对象
11      private ServiceConnection conn = new ServiceConnection() {
12          @Override
13          public void onServiceConnected(ComponentName name,IBinder
14  service) {
15              audioService = (AudioService.AudioBinder)service;
16              //绑定服务后需要服务执行的操作需要写在这里
17              audioService.startAudio(AudioActivity1.this,
18                      list.get(currentPosition).getRawId());
19              audioPlay.setImageResource(R.drawable.audiopause);
20          }
21
22          @Override
23          public void onServiceDisconnected(ComponentName name) {
24
25          }
26      };
27
28      @Override
29      protected void onCreate(Bundle savedInstanceState) {
30          super.onCreate(savedInstanceState);
31          setContentView(R.layout.activity_audio);
32          initView();
33          Intent intent = getIntent();
34          currentPosition = intent.getIntExtra("position",0);
35          list = (ArrayList <Audio>) intent.getSerializableExtra("data");
36          audioTitle.setText(list.get(currentPosition).getTitle());
37          audioContent.setText(list.get(currentPosition).getContent());
38          //绑定服务
39          serviceIntent = new Intent(this,AudioService.class);
40          bindService(serviceIntent,conn,BIND_AUTO_CREATE);
41      }
42      void initView() {
43          audioPlay = findViewById(R.id.iv_play_pause);
44          audioNext = findViewById(R.id.iv_next);
```

```java
45          audioPrevious = findViewById(R.id.iv_previous);
46          audioTitle = findViewById(R.id.tv_title);
47          audioContent = findViewById(R.id.tv_content);
48          audioPlay.setOnClickListener(new View.OnClickListener() {
49              @Override
50              public void onClick(View v) {
51                  //调用服务中的方法
52                  boolean isPause = audioService.pauseAudio();
53                  if(isPause){
54                      audioPlay.setImageResource(R.drawable.audioplay);
55                  }else{
56                      audioPlay.setImageResource(R.drawable.audiopause);
57                  }
58              }
59          });
60
61          audioNext.setOnClickListener(new View.OnClickListener() {
62              @Override
63              public void onClick(View v) {
64                  audioService.stopAudio();
65                  currentPosition = (currentPosition +1)% list.size();
66                  audioTitle.setText(list.get(currentPosition).getTitle());
67                  audioContent.setText(list.get(currentPosition).
68                                  getContent());
69                  //调用服务中的方法
70                  audioService.startAudio(AudioActivity1.this,
71                          list.get(currentPosition).getRawId());
72              }
73          });
74
75          audioPrevious.setOnClickListener(new View.OnClickListener() {
76              @Override
77              public void onClick(View view) {
78                  audioService.stopAudio();
79                  currentPosition = currentPosition -1;
80                  if(currentPosition == -1){
81                      currentPosition = list.size() -1;
82                  }
83                  audioTitle.setText(list.get(currentPosition).getTitle());
84                  audioContent.setText(list.get(currentPosition).
85                                  getContent());
86                  //调用服务中的方法
87                  audioService.startAudio(AudioActivity1.this,
```

```
88                      list.get(currentPosition).getRawId());
89              }
90          });
91      }
92
93      @Override
94      protected void onDestroy() {
95          super.onDestroy();
96          audioService.stopAudio();
97          unbindService(conn);
98      }
99  }
```

任务反思

编写并运行音频播放程序，将在代码编写及程序调试过程中出现的异常信息、产生原因及解决方法记录在下方。

问题1：_____

产生原因：_____

_____

解决方法：_____

_____

问题2：_____

产生原因：_____

_____

解决方法：_____

任务总结及巩固

小白：师兄，这个任务中学习的服务和广播使 Android 程序在开发过程中能够实现很多以前没能实现的新功能，如开机启动程序等。

师兄：说得没错。服务和广播的用处是很大的，尤其是服务可以使我们的程序在没有界面的情况下实现后台运行。

小白：任务中通过服务实现了在后台运行音频播放，这样就可以在运行其他程序的时候听音频了。

师兄：服务除了用来实现音频的后台播放以外，还可以实现很多其他的功能，例如：消息的推送功能，即在程序不在前台运行的情况下也可以接收到服务器推送过来的消息，这是很实用的功能。但是，根据 Android 系统的运行机制，会杀掉长时间不运行在前台的组件，小白能不能找到方法来防止这种情况的发生呢？

一、基础巩固

1. 下列方法中，（　　）可以用来启动服务。
   A. startService(　)　　　　　　　　B. stopService(　)
   C. unbindService(　)　　　　　　　D. bindService(　)
2. 在绑定服务时，（　　）起到了服务与组件的通信作用。
   A. Intent　　　　　　　　　　　　B. IBinder
   C. Context　　　　　　　　　　　D. Service
3. 在绑定服务时，需要调用（　　）方法将服务解绑。
   A. destroyService(　)　　　　　　　B. stopService(　)
   C. unbindService(　)　　　　　　　D. onDestroy(　)
4. 下列方法中，哪个方法可以用于动态注册广播接收者（　　）。
   A. registerReceiver(　)　　　　　　B. sendBroadcast(　)
   C. unregisterReceiver(　)　　　　　D. sendOrderedBroadcast(　)
5. 下列关于广播说法正确的是（　　）。
   A. 广播是 Android 中进行消息传递的一种有效机制
   B. sendBroadcast(　)可以用来发送有序广播
   C. 自从 Android 8.0 之后，很多广播都不能进行动态注册
   D. 无序广播是可以截断的

二、技术实践

1. 查找资料，尝试通过后台服务实现网络下载功能。
2. 利用广播实现强制下线功能。

用户成功登录后，跳转至新页面；在新页面中，单击按钮发送强制下线广播；接收到广播提示"重复登录"，并返回登录页面。

任务3　项目打包

任务描述

将"学习通关"项目打包为".apk"文件。

任务学习目标

通过本任务需达到以下目标：

➢ 能够完成"学习通关"项目的打包。

任务实施

项目开发完成之后，如果需要发布到互联网上供别人使用，就需要将自己的项目打包成为 Android 安装包文件，其扩展名为".apk"，也就是我们常说的 APK。下面我们具体学习打包的过程。

（1）在菜单栏中单击 Build→Generate Signed Bundle/APK，选择"APK"，单击"Next"按钮，如图 4 - 17 所示。

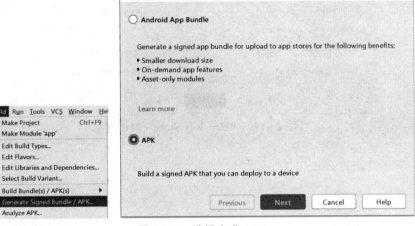

图 4 - 17　选择生成 APK

（2）弹出"Generate Signed Bundle or APK"对话框，如图 4 - 18 所示。

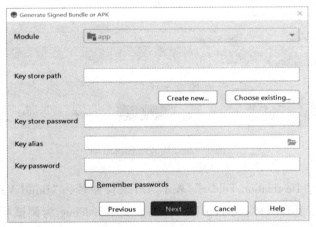

图 4 - 18　"Generate Signed Bundle or APK"对话框

其中，"Key store path"用于选择程序证书地址，由于是第一次开发程序，还没有证书，需要创建一个新的证书。单击"Create new"按钮，弹出"New Key Store"对话框。

如图 4 - 19 所示，信息填写完毕之后，单击"OK"按钮，返回到"Generate Signed

Bundle or APK"对话框。在对话框中，填写好路径及密码，如图4-20所示，单击"Next"按钮，选择构建类型及签名方式，如图4-21所示。

图4-19 "New Key Store"对话框

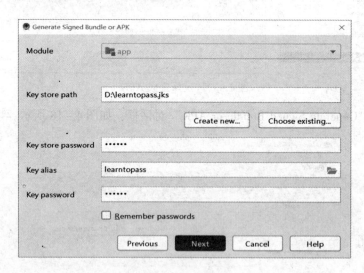

图4-20 填写路径及密码

在图4-21中，"Destination Folder"表示API文件路径，"Build Variants"表示构建类型，可选择的构建类型有两种：debug和release。其中，debug为调试版本，它包含调试信息，便于程序调试。release为发布版本，通常会做优化，以便用户更好地使用，此处选择release。"Signature Versions"表示项目打包时的两种签名方式：V1和V2。其中，V1表示传统通用方式，V2采用Android 7.0引入的新方式，安装更快，验证方式更安全，但不适用于旧Android版本，此处两个选项都勾选。

单击"Finish"按钮，开始构建。构建完成后，会在 Event Log 中提示如下信息，提示 APK 成功生成，如图 4-22 所示。

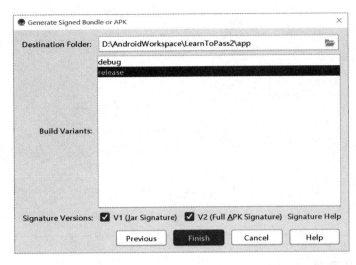

图 4-21 选择构建类型及签名方式

打开"D:\AndroidWorkspace\LearnToPass2\app\release"目录，可以看到打包好的 APK 文件，如图 4-23 所示。

至此，项目已经完成打包。打包成功的程序能够在 Android 手机上进行安装运行。

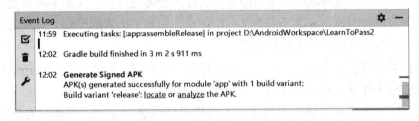

图 4-22 打包成功提示信息

图 4-23 生成的 APK 文件

任务反思

打包项目，将在打包中出现的异常信息、产生原因及解决方法记录在下方。

问题1：_____

产生原因：_____

_____

解决方法：_____

_____

问题2：_____

产生原因：_____

_____

解决方法：_____

任务总结及巩固

师兄：小白，到这里我们的"学习通关"项目开发就基本可以告一段落了。

小白：从第1个页面到现在已经完成了4个功能模块，一路走来遇到了很多问题，也感受了解决问题的喜悦，真是不容易。

师兄：软件开发就是这样一个有吸引力、让人着迷的过程。技术不断在更新，困难不断会遇到，但是当我们解决了一个又一个困难，积累了丰富的经验，开发出用户满意的产品时，这种成就感也是无可比拟的。

小白：学得多了，才发现自己不会的也很多。还有很多功能效果，我还不会，但是现在我有信心去各个攻破，让我的"学习通关"更加丰富。

师兄：IT技术发展的脚步太快了，在我们说话的这会工夫，可能又有旧的技术被淘汰、新的技术出现了。这就需要我们不断地学习、不断地实践，才能不被这技术大潮淘汰。小白，作为一个新手程序员，你未来的路还长着呢！希望你勤学不辍，成为一名合格的程序员。

请运用本书中学习的各项技能，继续完善"学习通关"各项功能。

学习目标达成度评价

了解自身对本模块的学习情况，完成表 4-3。

表 4-3 学习目标达成度评价表

序号	学习目标	学生自评
1	能够使用 VideoView 和 MediaPlayer 实现视频和音频的播放	□能够熟练实现 □通过查看以往代码及课本，能够基本实现 □不会实现
2	能够使用启动和绑定两种方法创建和启动后台服务	□能够熟练实现 □通过查看以往代码及课本，能够基本实现 □不会实现
3	能够创建和接收广播	□能够熟练实现 □通过查看以往代码及课本，能够基本实现 □不会实现
4	能够打包项目生成 APK 文件	□能够熟练打包项目 □通过查看课本，能够基本实现 □不会实现
评价得分		
学生自评得分（20%）	学习成果得分（60%）	学习过程得分（20%）	模块综合得分

（1）学生自评得分。

每个学习目标有 3 个选项，选择第 1 个选项得 100 分，选择第 2 个选项得 70 分，选择第 3 个选项得 50 分，学生自评得分为各项学习目标得分的平均分。

（2）学习成果得分。

教师根据学生阶段性测试及模块学习成果的完成情况酌情赋分，满分为 100 分。

（3）学习过程得分。

教师根据学生的其他学习过程表现，如到课情况、参与课程讨论等情况酌情赋分，满分为 100 分。